Living with Birds

Living with Birds

Asad Rahmani

juggernaut

JUGGERNAUT BOOKS
C-I-128, First Floor, Sangam Vihar, Near Holi Chowk,
New Delhi 110080, India

First published by Juggernaut Books 2024

Copyright © Asad Rahmani 2024

10 9 8 7 6 5 4 3 2 1

P-ISBN: 978-93-5345-415-9
E-ISBN: 978-93-5345-217-9

The views and opinions expressed in this book are the author's own. The facts contained herein were reported to be true as on the date of publication by the author to the publishers of the book, and the publishers are not in any way liable for their accuracy or veracity.

All rights reserved. No part of this publication may be reproduced, transmitted, or stored in a retrieval system in any form or by any means without the written permission of the publisher.

Typeset in Adobe Caslon Pro by R. Ajith Kumar, Noida

Printed at Replika Press Pvt. Ltd., India

Dedicated to my teachers, parents and family, students, colleagues and BNHS family.

Contents

1. Prelude — 1
2. The Early Years — 6
3. At Home in AMU — 29
4. Early Conservation Struggles — 43
5. Joining BNHS — 53
6. The Bird That Changed My Life — 57
7. Changes Come Calling — 78
8. The Florican Project — 88
9. Desert Tales — 97
10. AMU Revisited — 105
11. Back at BNHS — 120
12. The BirdLife Partnership — 137
13. The Spirit of BNHS — 144
14. 125 Years and Counting — 160
15. Across International Lines — 166
16. The Power of Partnership — 172
17. The Legendary Ali Hussain — 184
18. The Vulture Crisis — 193
19. Field Visits — 209

20. More Feathered Friends	222
21. For the Love of Students and Science	243
22. My BNHS Colleagues	260
23. Field Assistants	266
24. The Forest Job	279
25. Colleagues in Conservation	292
26. The 'Babu' as Conservationist	302
27. Moving on From BNHS	310
28. The Corbett Foundation	317
29. Looking Ahead	327
Notes	340
Acknowledgements	347
A Note on the Author	349
About Indian Pitta	351

1

Prelude

A small advertisement in the *Times of India* in 1980 altered the course of my life significantly. The Bombay Natural History Society (BNHS) was looking for young field biologists with a degree in zoology or botany for their research projects. The advertisement appeared on a day (16 February 1980) when much of India was gripped by the fear of the impact of a total solar eclipse. For a few weeks before, newspapers were replete with articles on the effect of a total eclipse; mostly alarming, with some bordering on hysteria. I would argue with friends in Aligarh, saying there was nothing to fear because this natural phenomenon had been happening since the solar system came into existence 4.5 billion years ago, and would go on till our solar system died, in another 7 billion years. However, the frenzy was so huge, particularly in local newspapers and radio broadcasts, that even those with a scientific mindset entertained apocalyptic thoughts. Some scientists rightly suggested avoiding outdoor activities during the eclipse and specifically warned against looking at the sun with naked eyes; however, it was safe to watch the sun through a filter made of black-and-white photographic film negatives. Since I had a camera, I had accumulated many negatives and I distributed some of these to my more rational friends, who were keen to view the rare event. But my objective at the time of the eclipse was not really to see the spectacle, but something entirely unrelated.

Accompanied by a young Abdul Jamil Urfi, who is now a professor in Delhi University, I decided to go to Sheikha Jheel to watch birds during the total solar eclipse. With the eclipse scheduled for 2.30 p.m. in Aligarh and northern India, we set out around 10 a.m. from the bus stand. Abdul had arrived at my hostel room (at that time, I lived in Aligarh Muslim University's [AMU] Habib Hall) early in the morning, brimming with excitement at the prospect of witnessing and documenting abnormal bird behaviours during the eclipse. Laden with sling bags weighed down by cameras, binoculars, notebooks and provisions, we pedalled towards the bus station on our cycles. An eerie silence greeted us; the usually crowded and noisy streets of Aligarh were empty. The shops were shut, cycle rickshaws were conspicuous by their absence, cars were off the roads and even the dogs were sulking in corners; it was only the presence of an occasional human being that confirmed that there was life in the city. When we reached the bus stand (where we were planning to park our cycles), idle buses seemed to tease us with their immobility. When I asked the lone manager, he said, '*Aaj koi bus nahi chalegi, sooraj grahan ke karan* (No bus will run today due to the solar eclipse).'

Undeterred, Abdul suggested that we cycle down to the Jheel, some 16 km away. Despite the age gap – he was 15 and I was double his age – our shared excitement propelled us forward. Determined to document our observations for the

Journal of the Bombay Natural History Society (JBNHS) – we envisioned significant findings on bird behaviour during the period of the solar eclipse – we pedalled through deserted roads, encountering only two villagers along the way, and eventually made our way to Sheikha Jheel.

We arrived at the Jheel around midday, and almost immediately immersed ourselves in its vibrant bird life. After a brief rest of about 20 minutes, and some nourishment, we decided to sit separately and note the behaviour of birds. The time of the eclipse arrived, and darkness slowly crept into the sky. The birds remained unperturbed by the darkening sun and continued their usual activities of foraging or resting. Seconds later, despite the brief period of total darkness, none exhibited any noteworthy strange behaviour – there were no sudden flights, unusual calls or any other remarkable reactions. What a contrast to some cowardly humans! I remembered my classmate (whom I had visited that morning), who was huddled up in his darkened room, wearing dark glasses bought specially 'to avoid the bad impact of the solar eclipse'. I decided then that living in the company of birds and wildlife was much more rewarding than engaging with such irrational people.

My rationality and scientific temper had been tested a year previously too, in 1979, when a huge ruckus was created by the media over the falling of debris from the disintegrating Skylab space station, a predecessor to the currently operational International Space Station. Everyone thought that the debris would fall on their head, but I maintained that the likelihood of it landing in Aligarh was one in a billion. One practising Muslim, knowing that I did not believe in God or prayers, conspiratorially advised me, '*Qayamat aane wali hai, ab to namaz padho*. (The world is going to end, at least start praying now).' In the end, some of the debris fell into the Indian Ocean, and other parts landed on land, in a remote part of western Australia! Fortunately, there was no *qayamat*. Until I left Aligarh in 1980, I would jest with that devout individual, asking him about the fate of his anticipated qayamat; the only response was a stupid grin.

That solar eclipse day in February 1980 marked a turning point in my life. While the sun disappeared for a few minutes that day, I could see bright light ahead, shining on my own future – I could be joining BNHS if my application in response to the advertisement published that day was successful. In 1973, after completing my MSc in Zoology at AMU, I had written to J.C. Daniel, the curator of BNHS at that time, and expressed my desire to pursue a PhD on birds under Dr Sálim Ali. He promptly wrote back, saying that while they did not have any scholarship or stipend for researchers, I was welcome to join BNHS. A few months later, I obtained a scholarship from the Department of Science and Technology, but the modest sum of Rs 300 per month was inadequate to support a life in Bombay (now Mumbai). I reluctantly dropped the idea and instead enrolled for a PhD in the Zoology Department of AMU to work on the olfactory organs of fish. Meanwhile, I continued to nurture my passion for wildlife through putative fieldwork (such as going to the University's Qila or Sheikha Jheel for bird watching), reading and writing. I remained in communication with J.C. Daniel and Dr Sálim Ali, and kept them updated on my activities. Later, in February 1980, when J.C. Daniel reviewed my application, he expressed his delight. I was called for an interview in March and the much-awaited appointment letter from the BNHS arrived in the mail in April.

Prelude

I start my memoirs with this prelude, as it brings together three very important themes of my life – the importance of a scientific temper, my interest in conservation and BNHS. These have mingled and defined my entire life and career; the pages that follow join these very large dots. How successful I have been in drawing the picture, I leave the reader to judge.

2

The Early Years

My grandfather worked for the postal department during the British period. He had three sons and a daughter; my father was his second son. One of my early memories of my grandfather is a rather unhappy one – he was weeping at the demise of his youngest son, who died young, five to six years after relocating to Pakistan during the Partition. Witnessing my grandfather's tears left a lasting impression on me. With his striking appearance including a long white beard, strong carriage, gentle voice and large headgear, he resembled Santa Claus. But did Santas cry? At the age of eight, I could not comprehend life's complexities.

Another lasting memory of my grandfather is his exquisite handwriting, which resembled calligraphy. He taught us, his grandchildren, the art of writing in the Urdu script, using a *takhti* (wooden plate) and a pen crafted from the stem of *sarkanda*, a native wild grass (*Saccharum munja* or *Saccharum bengalense*). The unique feature of the *takhti* was its reusability; after writing, it could be refreshed by applying a white paste made from *chikkni mitti* (wet clay soil), followed by a period of drying in the sun for a few minutes. I recall how my elder brother, Feroz, and younger sister, Nishat, and I would sing songs while waiting for our *takhti*s to dry, before proceeding to our next writing lesson.

Despite his modest salary, my grandfather ensured that his children received an education. My father, after studying law, eventually passed the judiciary exam and secured a position as a *munsif* (judge), with a grand salary of Rs 72 per month during the British Raj. I cannot recall the exact year, but my father often recounted this achievement later in life, emphasizing the value of money.

Our family consisted of seven siblings: Khalid, Mujahid, Naheed, Parveen, Feroz, myself and Nishat. No matter where my father was stationed, our parents ensured that we received top-notch education in the finest schools and colleges. All my sisters attained an MA degree before they got married (matrimony was entered into only with their full consent), while three of us (Mujahid, Feroz and myself) pursued and obtained PhDs. My eldest brother retired as a regional bank manager of the Union Bank in Lucknow. Unfortunately, three of my brothers passed away early, due to heart issues associated with smoking. Despite the prevalence of smoking among family and among college friends, I have never touched a cigarette in my life.

As a judge, my father was transferred to various districts every two to four years, and I grew up in places like Bijnor, Meerut, Saharanpur, Rampur, Budaun and Agra. At Saharanpur, where he served as the district judge, we lived in a spacious British-era *kothi* (a large house) with 11 rooms and a large compound; we did not have enough furniture to furnish all the rooms, resulting in four to five rooms being kept locked.

Saharanpur has a large cantonment, famous for its horse and mule training facilities. Whenever my father used to visit his army friend, Brigadier Bijli, I would tag along, primarily to see the young army officers training the horses. At that time, Dehradun and Mussoorie were also under Saharanpur district, and for almost two months in the summer, my father's *kacheri* (court) would shift to Mussoorie, to a building just opposite the famous Hakman's Hotel on the Mall Road. We enjoyed our summer vacations in Mussoorie for three consecutive years. Later in 1967, my elder brother, Mujahid, just after completing his MSc in Chemistry, taught there for a year, giving me an opportunity to go back to Mussoorie for another summer. Some years later, my eldest brother, Khalid, was posted as the bank manager of the Union Bank in Mussoorie for three years. His office-cum-residence was situated near the iconic clock tower in Landour. Consequently, for almost nine years, I spent my summer holidays in Mussoorie, creating cherished memories amid its beautiful landscapes.

Besides the Camel's Back Road, adorned with large pine trees that provided solitude, my other favourite place was a *raddiwalla* shop, from where I would purchase comics for 4 annas (25 paisa), and old books for Rs 1 or 2 per book. Pocket money of Rs 10 a month was insufficient to fulfil my appetite for comics and books. To supplement my reading, I started spending time in the Mussoorie Library, which had a large collection of books. The bonus was a magnificent view of the valley from its many windows.

From Saharanpur we shifted to Rampur, where I completed high school from Raza Intermediate School. I was 15 by this time, and the challenges of growing up and surges of testosterone fuelled a suppressed teenage rebellion in my mind and body. My leftist ideology and atheism often led to heated arguments with friends, all of whom were either devout Muslims or Hindus. Rampur was home to an ageing nawab who would invite my father for discussions, as his own son had rebelled against him. Not taking any sides, my father would also meet the nawab's son, Zulfiquar Ali Khan (aka Mikki Mia). At our dinner table, my father would often share stories of the nawab family's machinations – revolving around the pursuit of family properties and wealth. Once, the nawab invited our whole family for dinner, but I refused to go due to my leftist beliefs and leanings; how could I dine in a bourgeoisie house?

But these feelings were attenuated by my surroundings. In Rampur, I would see huge Mughal-type buildings inside the Rampur Qila, the nawab's old palace, along with numerous expansive properties, wide roads and towering gates dating back to the nawabi era. However, beyond this historical splendour, the city itself was engulfed in chaos: broken roads, overflowing drains, poorly constructed and dilapidating houses, garish shops and messy traffic – the usual ills that characterizes urban India. The only building that I liked in that chaotic city was the famous Raza Library, which houses thousands of rare books and original manuscripts. In my view, the rulers of the erstwhile Rampur estate redeemed themselves through this priceless legacy that stood out as a beacon of culture and knowledge amid the city's muddled surroundings.

After three years in Rampur, in 1967 my father was posted to Budaun, another unassuming medieval town with Mughal-era buildings and a fort, now in ruins, and a beautiful clock tower known as the Ghanta Ghar. Badaun is renowned for its specialty sweet, the *Badauni pedha*, which can be stored for extended periods, making it a favourite among travellers and an ideal gift item. As our large house was situated on the outskirts of the town, I developed a keen interest in trees, agricultural fields and a lazy meandering river that served as my sanctuary for birdwatching. It was much later that I came across literature mentioning the presence of Wild Buffalo (*Bubalus arnee*) during the Mughal period in the Oudh region, particularly in the flood plains of Budaun. By the late 1960s, the only large wild animals left were the Nilgai (*Boselephus tragocamelus*) and scattered herds of Blackbuck (*Antelope cervicapra*). Unfortunately, I am not aware of the current status of the natural areas of that region and whether any of these species have been preserved in the landscape.

The 1960s was the age of the hippie movement, which emerged as a rejection of the mores of mainstream American culture. With their long hair, tacky clothes, maverick attitude, songs, free sex and drug use, hippies became emblematic of a rejectionist ethos towards societal norms. Reading about hippies in newspapers and magazines fuelled a long latent rebellion in a relatively uneventful small town like Budaun, where nothing interesting ever happened. Growing shoulder-length hair, much to the dismay of my father and the shock of the conservative citizens of Budaun, was my way of expressing dissent against the traditional values prevalent in the Indian society. I desisted (much to their relief!) from drug use or other aspects associated with the hippie lifestyle. However, my hirsute uprising was short-lived – a few months later, I returned to my studies with a more conventional appearance.

Around this time, two world events solidified my love for science. The first was the historic human-to-human heart transplant performed on 3 December 1967 in South Africa, which made front-page news. I remember seeing, most likely in *Life* magazine, the pictures of Louis Washkansky, the patient, and a team of doctors led by Dr Christian Barnard. The second event was the moon landing on 16 July 1969. For many weeks, newspaper columns were full of pictures – there were full-page photographs of Neil Armstrong and Edwin Eugene 'Buzz' Aldrin walking on the moon's surface in their bulky space suits and backpacks of oxygen.

I've loved animals since childhood and had many pets. When I was six or seven years old, and living in Meerut, I rescued a parakeet chick that had fallen out of an old neem tree in the compound and nursed it back to health. My mother healed its injuries with *haldi* paste, and in a few weeks it transformed from a gawky chick with down feathers into a robust bird, ready to bite the fingers of my siblings (as its main caregiver, I had learnt to handle it). Fruits brought by my father were shared with it, but sadly, my lovely Rose-ringed Parakeet wasted more food than it ate; I was unaware at that time that this is a common habit among all parakeets. Much later, I realized that there is no wastage in nature. Small fruit remnants that parakeets and parrots drop on the ground become accessible food for ants, rodents, squirrels, birds and other animals. Langurs are also messy eaters and drop scores of leaves and fruits, while cheetals wait below to feed on them. I saw this lovely symbiotic relationship in Kanha a few years ago.

We had many *murgis* (chickens) of various kinds, such as Rhode Island Red, Aseel, Nicobari and Leghorns, in those large old houses allotted to judges. Cleaning their pens, feeding them grain and running after them with my younger sister Nishat was a regular pastime after school hours. My pet *tota* (parakeet) became friendly with one large cock and would rarely be seen away from him, sometimes even taking a ride on his back. When the cock became sick, the parakeet did not leave his side. Unfortunately, after a few days, the cock died, leaving my beloved *tota* in distress – it refused to eat for some days. Such unexpected animal behaviour left a deep impression on my young mind.

In North India (and perhaps elsewhere too), it is a common maxim that even if one keeps a parakeet for *das saal* (10 years), as soon as it finds its way from the cage, it will escape and never come back. The Urdu word for selfish people is *totachasm*. But my *tota* (that was never caged in the first place) did not leave me till his natural death. How wrong are we in our ideas about the animal world!

After a three-year posting in Meerut, my father, a session judge at that time, was transferred to Tehri Garhwal. Some called it a punishment posting (for being too honest!) but my father took it sportingly, being the adventurous person that he was. At that time (the late 1950s), Tehri Garhwal did not have good schools and colleges, and so he decided to station the family at Meerut. His jurisdiction covered almost the whole of Garhwal, Ranikhet, Lansdowne, Pauri, Chamoli, Rudra Prayag and Uttarkashi. Based in Tehri Garhwal, my father would go to other districts with his *kacheri* because, at that time, the government found it more convenient to send the court from town to town, instead of requiring people to come to the district headquarters in Tehri Garhwal. 'Justice will come to your door' was the maxim then. Nice to hear this in present-day India!

My father was fond of nature, *shairi*, adventure and driving, and despite his family being so far away, retained a passion for his job. We children would wait for summer and winter holidays to visit him. In Tehri Garhwal, I had a pet Chukar Partridge (*Alectoris chukar*) for a few months, and it followed me around the house like a pet dog. I do not remember what happened to it when I left for Meerut after the summer holidays. In 1958, my father brought home two Alsatian pups, a male and a female, given by the erstwhile maharaja of Tehri Garhwal. For the next 14 years, till their death due to old age, Rex and Rani, as we had named them, were my life. We were 12 family members now – father, mother, grandfather and seven siblings, and Rex and Rani. All my free time was spent with these beloved dogs. I never treated them as mere animals – they were a part of my existence. Feeding, grooming and bathing them, and playing and running with Rex and Rani were my after-school activities. The affections were reciprocated; they would not eat until I fed them, and as soon as I returned from school it was playtime for the three of us. In the hot summer months of North India, they would sleep beneath my cot in the air-cooled room. Even when I fell sick with chicken pox and was isolated, Rex and Rani did not leave me alone; they truly taught me the value of love, affection and caring. When their end came in 1972 (both passed within a gap of three days due to old age), I was inconsolable for a few days. Now, 52 years after their demise, I still miss them.

In addition to Rex and Rani, I had other exotic pets – my father once brought down a young Himalayan Brown Goral (*Nemorhaedus goral*) from Garhwal to Meerut. At that time, I was unaware of its distribution, and was told that it lives in the Himalayas and was called a *pahari bakri* (mountain goat).

We named it Bambi and it went on to survive the heat and dust of the plains, buoyed by indoor rest in the air-cooled room (facilitated by the air cooler) during the day, feeding on grass and kitchen vegetables for three years. Bambi delighted our guests, and some came specially to see it. One 'uncle' was thoroughly disappointed after seeing my Bambi. 'This is simply a *pahari bakri*. Its horns are not even impressive,' he said. I never forgave him for his rude comment about my charming pet.

Over the years, some of my other pets included pigeons, chickens, peafowl, lovebirds, parakeets, munias, a hill myna, spiders, fish, a toad, a hedgehog, grey francolins, a praying mantis, bulbuls, rabbits, guinea pigs, beetles, moths, a turtle and any 'abandoned' animal that I found. Unfortunately, many of them did not survive for long but I made every attempt to feed and care for them, aided merely by my incipient knowledge of wildlife. Watching them closely became a habit that helped me later in life, when I was studying wildlife in the field. Each pet taught me something – for instance, through their constant swimming within a tank, fish made me aware of endless hours of energy, while the toad taught me how to look upon the world placidly.

Each pet also brought home a different lesson in natural history and animal behaviour. We had several kinds of pigeons – fantails, pouters, white and homing. They all bred vigorously, like pigeons do, one clutch after another; even before the young had left the nest, another set of two eggs were laid. Feeding and housing the growing brood was becoming difficult, so I started making omelettes with the pigeon eggs, and sometimes replaced the eggs that were removed with *murgi* eggs. It was a challenge for the pigeons to incubate such a large egg, but they still readily accepted them. The precocial chick would emerge after 21–22 days, and the 'foster parents' would try to feed the confused chick pigeon milk. *Encyclopaedia Brittanica* helped me learn the difference between precocial and altricial chicks, and the strange thing known as pigeon milk. I remember how once a pigeon mother was trying to feed a hen chick the 'milk' oozing out of her mouth; she kept chasing the precocial chick, but it was busy running around, picking up bajra grains (millet) that I had spread! Hilarious as these episodes were, they also taught me about the innate behaviour of animals.

My adventures continued in school; at St Mary's School in Meerut, while in grade two, I became fascinated with the ability of Black Kites (*Milvus migrans*) to time their arrival with our recess, when we children would take our tiffin boxes to the large playground, particularly in the winter season. A dive-bomb from a kite and a piece of *paratha* or bread was gone from the hand, triggering laughter from the other children. It was amusing as long as it was someone else's food but I was the victim many a time. One of the games we played on the kites was to tie a white paper to one end of a string, while its other end contained a morsel of food. It was then thrown in the air for the kites. Dexterously catching the food, the kites would eat it on the wing, with the white paper whirling behind its tail. I often wondered how the kites, not seen otherwise during school hours, would magically appear in large numbers just before the recess to snatch food from an unwary kid. Did they listen to the bell announcing the recess, or follow the arrival of students into the open; or did they follow the sun? The first didn't seem probable – the sound of the bell did not travel very far, and St Mary's School was huge, with a large playground; so, it seemed unlikely that kites, which were seemingly far away from the school, would hear the bell. If they flew in after seeing the students move into the open, why did they not do so when we had physical training classes or when we played football? And finally, did they follow the sun and know that midday signalled lunch time? At that age (and now too!), my mind was full of questions.

In Budaun district, where my father was posted from 1967 to 1969, we had a large, colonial-era house at the edge of the town. In a row of nearly 800 m were the *kothi*s of the district magistrate, chief medical officer, superintendent of police and the district judge. All British-era *kothi*s were surrounded by large trees and bushes – resembling a veritable jungle. The huge lawns and gardens supported many birds and small mammals. Agricutural fields stood right in front of our houses and a forgotten *mazar* (grave) about a kilometre away. After college, I would spend endless hours roaming in the agricultural fields with my dogs, observing the marvels of nature, such as the endless stream of carpenter ants collecting fallen grass seeds. The dogs were in their element, chasing terrified hares (never killing them though), or flushing a covey of grey francolins, which would scatter, scolding the dogs as they flew in different directions. Further down, about 2 km from our house, was a slow, winding river, where I saw a pair of Brahminy Duck (*Surkhab* in Urdu) for the first time. I had heard of the Urdu metaphor '*Surkhab ke per lagey hain kya*' ('Are you so special?'), for it is difficult to obtain their orange-brown feathers. When I saw the Brahminy Duck breeding in Ladakh in 2005, my first sighting in Budaun flashed in front of my eyes.

One day, from my room window, I noticed a boy running away with three peafowl eggs. I chased him and 'rescued' all the eggs. Since he had seen the nest, it was not possible to prevent future raids, so I decided to keep them under my bantam fowl. All the three eggs hatched – two males and one female. Being precocial, they soon started following their foster mother, who would scratch rotting leaves to expose the hidden grubs, worms and insects. But soon it became too much for the poor hen to feed the three hungry peafowl chicks, besides her own five chicks. Since the peafowl chicks had to be supplied with insects, I would take them to our large garden, where they fed on their own, and I watched over them like a shepherd. I had to keep a close watch, as there were many mongooses around. While monitoring them, one day I noticed the incipient courtship behaviour of one chick (which later turned out to be a male) at the age of six days. The chick cocked its tail and moved around another peafowl chick (sex unknown), the way adult peacocks do. This observation found its way into a 1990 submission for the *Newsletter for Birdwatchers*.[1]

I also had a fish tank in my room, which I filled with small fish caught from the river. Along with the fish, I also scooped up some menacing-looking larvae. I did not know what species they were but nonetheless kept them in the tank, where they were bottom-dwellers. I had made my tank as natural as possible, with sand, pebbles, submerged plants and surface-floating vegetation. Maintaining it was an early lesson in ecology. Soon, I realized that some minnows were missing. At first, I thought that I had made a counting error, but skeletal remains made me suspicious. One evening, a larva with a helpless minnow in its mouth revealed the dark events that took place at night. As my aquarium was kept near the window, with direct access to sunlight for half the day, the wily larvae would lie hidden in the leaves and pebbles, waiting for darkness to reveal their true colours. I became fascinated by these creatures and would replenish the small fish, so that the larvae did not go hungry. After a few months, I found a dragonfly flying in the room – another natural history lesson learnt! One of the menacing-looking larva had metamorphized into an adult dragonfly. Many decades later, when I went dragonfly watching with Dheerindera Singh of Agra, an amateur naturalist, who has written a wonderful book on these insect predators, I told him of my childhood experience. He laughed and said, 'Sir, dragonflies are the tigers of the insect world.'

These growing-up experiences fostered a deep love for biology, a subject I wished to pursue in grade 12. But having scored 94 per cent in mathematics in high school, my father wanted me to become an engineer. Being a judge and someone with a domineering personality, my father sometimes forgot that he was dealing with his son, not a culprit in the court! I rebelled against his wishes, and declined to study mathematics, physics and chemistry in grade 12, flatly refusing to go to the school in Budaun. I lost a year in this fight, after which my father finally agreed that I could shift from the physics stream to biology. Not having studied the subject in grade 11, a biology teacher was appointed to tutor me at home. This retired teacher had taught biology for 40 years, and his concepts were clear. And so, thanks to his teaching and my interest in the subject, I was soon hurdling over concepts such as animal anatomy, physiology, classification and ecology. I topped the school in biology in grade 12. Not surprisingly, pursuing a career in biology (wildlife science had not been heard of in small-town Uttar Pradesh [UP] in 1968!) became the mission of my life.

After Budaun, my father was stationed in Agra, from

where, three years later, he retired as a district judge. At Agra, I completed my BSc degree from R.B.S. College, studying zoology, botany and chemistry, all the while making it clear to my parents that I had no intention of pursuing a career in medicine (in those days, anyone studying these subjects was expected to compete in the MBBS track). By the end of my BSc studies, I developed a strong desire to become a wildlife biologist, a field that was relatively unheard of at the time. Influenced by the works of personalities such as M. Krishnan, E.P. Gee, Jim Corbett and Sálim Ali, I felt inspired to pursue this unconventional path.

Meanwhile, the rest of the family was making strides of its own, albeit along more conventional lines. My siblings married and raised families. As far as I was concerned, after securing a job at the BNHS in 1980, I became deeply engrossed in fieldwork, leaving little time to contemplate marriage. Moreover, with my limited salary at that time, I could not afford a house in Bombay. My work kept me significantly occupied, taking me to various locations – from Karera in Madhya Pradesh (MP) to Nannaj in Maharashtra to Rollapadu in Andhra Pradesh, as well as to Rajasthan, UP, Assam, Gujarat and beyond. I do not know exactly when the age for marriage passed me by. My 'marriage' to my work was all-consuming, and the love and affection of my colleagues and students fulfilled my emotional requirements. In retrospect, was that a mistake? Maybe, but at that time it seemed like the right choice.

None of these doubts, however, take away the great satisfaction I feel when I witness the success and happiness of my nephews and nieces, who are thriving in happy marriages with children of their own. A vicarious pleasure may not be a substitute for the real, but if you have a large happy family (scattered all over the world as they are), what difference does it make? I am proud of them, and they are proud of me. This mutual sense of pride and happiness within our extended family is truly invaluable.

Religion was a big part of my life growing up – my father was a devout Muslim, offering *namaz* five times a day. We had many Islamic books, both in Urdu and English, that he would encourage us to read. At the same time, we also had copies of the Bhagavad Gita, Ramayana, the Bible and the Guru Granth Sahib. Often during dinner, he would discuss various religions – not the differences but the similarities. He also told us to respect every religion equally, while following our own. He would often point out that the Quran says that God has sent prophets and messengers to every part of the world, and it is only we who have forgotten their true message, indulging only in rituals. For us, Ram was *Imam-e-Hind* (prophet or religious leader of Hindustan) as the great philosopher-poet Allama Iqbal called Lord Ram. My father told us about the amalgamation of the Bhakti movement (which started in South India in the eighth century and slowly spread northwards) and the Sufi movement (which came much later), explaining that both call for love, affection, compassion, empathy and understanding of different faiths, and how both propagate the universal message of devotion to one common God (or Brahma). We were told that the Sufi and Bhakti movements, which inspired and articulated social messages, almost merged through poetry, songs and music.

We also learnt the basic similarities between the Upanishads and Quran.

These early teachings were influential, and could possibly explain why all our family members had friends, 'uncles' and 'aunties' from every community. I also believe that time and space cannot separate people if they have mutual respect for each other's religion and culture. Most of my best friends are Hindus, and some of them, such as B.C. Choudhury, Neeraj Srivastava, Raghunandan Chundawat, Dhritiman Mukherjee and Manoj Kulshreshtha, are like my brothers. Though a total atheist, I am greatly impressed by Hindu, Sikh and Buddhist philosophies.

Growing up, the time not spent with pets and family was devoted to books. My father was very fond of reading and introduced us to *Reader's Digest*, *The Illustrated Weekly of India*, *Life* and other magazines. As my elder siblings were interested in different subjects, such as sports, politics and literature, we used to get many types of magazines at home; in all around 18–20 publications would reach our home.

My favourite books from high school days onwards included *The Naked Ape* by Desmond Morris, *Fatu-Hiva – Back to Nature* by Thor Heyerdahl, *Coming of Age in Samoa* by Margaret Mead, *Silent Spring* by Rachel Carson, *The Cosmos Connection* by Carl Sagan, *Serengeti Shall Not Die* by Bernard Grzimek, *Born Free* by Joy Adamson and *The Deer and the Tiger* by George Schaller. In later years, the other books I enjoyed included *Autobiography of an Unknown Indian* and *The Continent of Circe* by Nirad C. Choudhury, *My God Died*

Young by Sashti Brata and *An Area of Darkness* by V.S. Naipaul. Ruskin Bond was also another favourite author.

Long summer holidays followed the annual exams, which were over by the end of April or early May. The two-month-long summer vacation, when one was confined indoors to avoid heat stroke, gave me ample opportunities to read. For us siblings, it was bonhomie time, with occasional fights for our favourite magazines! Our summer holidays were like Kashmir's 40 days of *Chalia Kalan* of yore, a period of intense cold when outdoor activities were suspended due to the freezing weather and 6–7 feet of snow. Though it kept Kashmiri families inside their houses, it gave them opportunities to interact with each other, learn cooking, read, write, indulge in philosophical discussions with elders and develop skills of weaving. Just as many of my Kashmiri friends miss the period of *Chalia Kalan* (as very few now stay home for long periods, due to modern infrastructure and heating facilities), I too miss the childhood period of summer holidays that gave us opportunities to read and interact with our brothers and sisters. These days, children are under intense pressure throughout the year to read textbooks, appear in exams and score high marks. Parents, too, do not have time for children. Another biggest reason for family disconnect is the mobile phone. Even within a house, siblings now 'talk' through the mobile. Who has the time to play ludo or carrom on a bed, and fight over who won?

Besides the numerous magazines that we used to subscribe to, I became addicted to cowboy western novels set in the American Wild West of the eighteenth century, dealing with settlers, outlaws and cowboys. The cover of those novels would be rather attractive, often featuring a handsome man on a horse wearing a Stetson hat with a broad rim, flashy shirt, flared jeans, a large leather belt with a fancy buckle and slant-heeled boots with spurs. For a testosterone-addled teenager craving 'action', these stories about the rebellious lives of cowboys in the vast Badlands of America were like a magnet. Echoes of that passion are reflected in the stylish hats I wore later in life (though I never learnt to ride a horse) – a small consolation to my large boyish dreams!

Lucknow is known for its *nazakat* (elegance), *nefasat* (nicety), *sheerin zuban* (sweet language), *shayarana shauq* (love for poetry), culinary delights and majestic nawabi-era buildings. What is lesser known are its libraries. While small libraries and shops of the Nakhas area (old Lucknow), where Urdu and Hindi books were sold, have largely shut down, the Amir-ud-Daula Public Library survives in all its magnificence. Built in 1868, it was initially a part of the government museum, but in 1887, it was opened to the public as a library. The whole precinct around the Library once had gardens (the beautifully designed Kaiserbagh Circle) and majestic buildings. The ravages of time and hedonistic developments have destroyed the Kaiserbagh Circle, but the Amir-ud-Daula Public Library stands alone in its splendid glory. Fortunately, the government has also given attention to its maintenance. When I used to visit it in the 1970s, its dark, cavern-shaped interior would not leave a good impression. Nonetheless, its rich collection of books greets one with warmth, the way only classical books can. After retirement, my father became chairman of the Sunni Wakf Board and the household shifted to Lucknow. Whenever I came to Lucknow from Aligarh during summer holidays in the early 1970s, visiting Amir-ud-Daula Public Library was the best way to spend time. Though, opening a copy of the *Encyclopaedia Brittanica* and noticing white silverfish (*Lepisma saccharina*) escaping from the spine could be irritating initially, the information on its pages would take you to the world of knowledge – leaving behind the world of pestilential insects!

Another favourite haunt was the Acharya Narendra Dev Library on the bank of the Gomti. Established by the former Chief Minister (CM) of UP, Mr Chandra Bhanu Gupta, in 1959 in memory of the great social reformer and thinker, Acharya Narendra Dev, it had good natural light, tall windows and spacious sitting areas. As it was established mainly for students – the Lucknow University is just across the river – the library used to subscribe to many scientific journals and the latest books. I would read journals such as *Scientific American*, *Science*, *Nature*; my old notebooks are testimony to the wealth of information I gathered from this library 50–55 years ago.

The third library that I would often visit was the British Council Library in the Mayfair Building in tony Hazratganj, a fashionable area of Lucknow. Mayfair housed the famous cinema hall of the same name, and it screened English movies; so, there was a double advantage in going to the Mayfair Building – the first being a chance to feast the brain for a few hours and then feast the eyes for two more hours.

Mayfair cinema hall suddenly closed in 1996, and in 2001 the British Council Library gave away all its books to Acharya Narendra Dev Library and wound up. For me, the only attraction left in that part of town was Advani's Book Store at the corner of Mayfair. In the 1980s and 1990s, whenever I would go to Lucknow from Bombay or Aligarh, a visit to Advani's Book Shop was a must. The shop's owner, Ram Advani, was a voracious reader on all subjects and would lovingly recommend books to visitors. He had a big fan following in the literary circles of Lucknow. The death of Ram Advani on 9 March 2016 (at the age of 95) led to the closure of the iconic bookstore; it had served the Lucknow intelligentsia for 71 years. For me, the tragedy is not the crumbling of bricks and mortar of the iconic Mayfair building, but the crumbling of reading habits in India. Imagine the closing of a fine bookstore in the most famous area of Lucknow. What does that say about us?

As I wrote earlier, my father was fond of reading good articles, and *The Statesman*, brought out from Calcutta (now Kolkata) and New Delhi, was his gold standard for quality writing. He would encourage us to read the newspaper for the high quality of its language. It was within the pages of *The Statesman* that I first encountered M. Krishnan's fortnightly column 'Country Notebook', which he started writing in the year I was born, 1950. I would sit with my father's Oxford Dictionary to understand Krishnan's articles. He was not only a top-class photographer but also a master craftsman of words and phrases. I would eagerly wait for his articles – the two-week gap between pieces seemed too long a time to wait! Krishnan's articles not only taught me natural history but also the beauty of the English language. I would dream of seeing the animals that Krishnan wrote about and of travelling to places such as Kaziranga, Corbett, Kanha and Bharatpur. Who knew then that my dream would come true? Although

I could never write like Krishnan, I had the opportunity to visit most of the places that my idol described. Krishnan's persona and work began to dominate me – not just my dreams but also my life's mission; like him, I wanted to be a traveller, writer, photographer, naturalist, conservationist, observer and humanist. That ambition is still a work-in-progress.

Another person who greatly influenced me in those early years was Edward Pritchard Gee, a British tea planter based in Assam, who was keenly interested in wildlife conservation. His masterpiece, *The Wildlife of India*, was published in 1962, and since then it has remained a favoured book companion; I must have read the book five to six times, if not more. Jim Corbett's books, which were part of my father's personal library, were also another favourite.

In the mid-1960s, I got my hands on Dr Sálim Ali's classic book *The Birds of India,* having come across it in my school library. That was my first 'encounter' with Dr Ali and an organization named the Bombay Natural History Society – I did not know then that both would change my life.

Along with reading, I developed the habit of asking questions, right from childhood. How does the quick transformation of summer-parched land take place with the first showers? Where does my 'friend', the large toad, a regular visitor to the veranda, hopping under the bulb (the perfect spot to gobble insects) right through summer and monsoon, disappear to in winter and why? Why do insects converge around a light where wall lizards wait to pounce on them? How does underground water come up when we use the hand pump? How can prayers and amulets make people healthy? If street mendicants can cure skin diseases, why do we need doctors? Questions, questions and more questions. Some were answered by my elders, but most were brushed aside.

One day at the dinner table, I asked my mother, 'From where do babies come?' My father replied, 'They come after marriage, *nikah*.' I responded with, 'There is no *nikah* in Hindus, how do they get babies?' leaving my parents and elder siblings nonplussed. A soft giggle from my eldest sister (a teenager at that time) followed, and then there was silence till we finished dinner. In the late 1950s and 1960s, there was no provision for sex education in schools. It was a taboo word, both at home and in school. Friends, and maybe househelp, were the sources (mostly unreliable) of information. When I saw my dogs having sex, I wanted to separate them, not knowing why they were entwined.

In my teenage years, I started questioning religious teachings and traditions. Why do I have to fast? Why should I fast for only 30 days, and not 20 or 40 days? Why do we have to face Mecca to say prayers? Why do I have to pray five times a day, why not just once a day? My father, a pious Muslim, tried to explain but I was never satisfied. As a common ruse, he would tell me to believe whatever is written in the Quran, ending with a final authoritarian statement, 'Everything cannot be explained!' This inane sentence would create more curiosity and questions in my mind. By the time I was 15 or 16, I became a non-believer in God. A few years later, I got hold of Charles Darwin's *On the Origin of Species by Means of Natural Selection*, which further cemented my conviction that God was a creation of human beings, and that we are not created by God. I also became interested in astronomy and geography – the two subjects that tell us about the vastness of the universe, Earth's miniscule position in the solar system and the beauty of our planet. If we are God's chosen species, why does the entire universe not revolve around Earth, as some semantic religions indeed teach? Over a period, asking questions became a habit, one that was sometimes irritating to family and friends.

Later in life, I would encourage my students to ask questions. Science grows only through curiosity, by asking questions and by trying to find answers. Now, at the age of 74, I still find myself asking questions and seeking answers!

3

At Home in AMU

In 1971, I applied to and secured admission to the masters' programme offered by the Zoology Department of AMU. As I was an 'outsider', having done my graduation elsewhere, there was much curiosity among the 'internal' students (those who had done their BSc from AMU and were now pursuing their master's) about me. Aligarh Muslim University was famous for its 'introduction rituals', an euphemism for ragging, that new students were subjected to. So, I felt some amount of trepidation attending the class for the first time. We had all gathered in the corridor; the 'internal' students, who had done their BSc from AMU were talking to each other, and some were meeting up after the summer break. I was alone, with just two or three 'external' students standing nearby. We knew the 'internal' students were planning to rag the new students. Our class was in front of the famous museum of the Zoology Department, where a large number of animal specimens were displayed, either mounted or in formalin jars. The museum is well kept, with proper labels and specimens displayed according to taxonomic classification. A large specimen of an Orangutan in a glass cabinet welcomes the visitor at the museum's entrance. However, the label that said 'Distribution: Africa' was wrong. So, I said out aloud, 'The label is wrong. The Orangutan is not found in Africa.' Another student, clearly the class bully, was put off by my comment. He pushed me and said, 'How can you say this? The label has been there for many years.' But I was firm in my reply, 'Even if the label has been there for many years, it does not mean that it is correct.' An argument ensued, bringing out the Head of the Department from his office. He thought that ragging had commenced even before the start of the first lecture. When I told the Head that a label was wrong, he was taken aback by this assertion from a new student. He must have seen the conspicuous specimen – the largest in the museum – hundreds of times, without noticing the error. He called the curator. I was called to the Head's chamber along with the curator; a few students accompanied us, and an encyclopaedia was hurriedly brought from the library. The curator admitted that the Orangutan was found only in Malaysia and Indonesia, and not in Africa. We were asked to go back to the class. The next day, a new label was put up, rectifying the previous mistake. The chastised bully became my friend for the rest of my stay in Aligarh.

The Zoology Department is one of the finest departments of AMU, home to inspiring teachers, such as Prof. Mirza, Prof. Zahoor Qasim (who later established the National Institute of Oceanography in Goa), Prof. Shah Masood Alam, Prof. Man Mohan (both famous entomologists), Prof. Nawab Khan (entomologist and toxicologist), Prof. Shamim Jairajpuri (a world authority on nematodes) and many more. They were all excellent teachers and guides, fostering students who excelled in their fields. I thoroughly enjoyed the two years of my MSc course. Prof. Musavi taught us evolution in such a powerful way that even diehard 'creationists' would change their views. But some of my fellow devout Muslim students continued to say, 'All this is okay, but what is written in the Quran is correct.' Can anyone change a person who blindly believes in God?

In the Zoology Library, along with textbooks and numerous journals, hidden in a corner I found bound volumes of the *JBNHS* dating back to 1906. Over the next five years, I read all the volumes that were in the library, making extensive notes on the natural history subjects. The *JBNHS* introduced me to A.O. Hume, C.H.T. Marshal, R.I. Pocock, E.C. Stuart Baker, R.G. Burton, Sálim Ali, Humayun Abdulali, Charles MacCann and others. What I could not study in the field at that time, I found in the venerated pages of the *JBNHS*.

In 1971, when I secured admission to AMU, Habib Hall had just been opened. It was a sprawling 300-room student residence, with a large dining room and an equally large common room. The lawns too were large enough to play cricket in. The provost wanted only good students in this prestigious Hall, and with my grades – always in the first class band, all throughout college – I was allotted a single room. For the next eight years, as a PhD programme followed the master's, 130-B, Habib Hall was my home.

The Cultural Education Centre of AMU, also called Kennedy Hall, has a large auditorium, where cultural events, such as *mushiaras* and *kavi sammelans* (poetry recitals), lectures by famous personalities, film shows, dramas and music festivals are held. I was a frequent visitor to the Centre and spent many evenings admiring M.F. Husain's mural, and the well-kept garden outside the Centre. Sitting silently in the garden, pretending to read, I used to observe the nesting of the Purple Sunbird (*Necterinia asiatica*) in a Bougainvillea bush. Watching them over many weeks, I saw the entire process – from nest building to chick fledging. My frequent presence triggered the curiosity of the members of the Drama and Literary Clubs, which had offices in Kennedy Hall. When I showed them the feeding of chicks by parent sunbirds, one student said, 'We pass this bush every day, but did not notice the pretty bird.' Later, everyone became custodians and protectors of the little flying jewels' nest. My friendship with the club members blossomed as the sunbird chicks fledged and left the nest. The next year, the sunbirds nested at almost the same place, hiding in plain sight in a place teeming with numerous visitors. My friend, Anees Ahmad Khan, a fine actor and the 'life' of the Drama Club, was the first to tell me about the return of the sunbirds.

The film shows in the Kennedy Auditorium were extremely popular, primarily because they were free. It had a seating capacity of 1,375 people, but often, we would get twice the number of students. Javed Lodhi, the secretary of the Film Club, once told me that the next movie to be screened would be *Ben-Hur*, the all-time classic. Anticipating an unmanageable crowd, we announced that a movie on fish would be shown! My PhD supervisor, an ichnologist, Dr Sardar Mehmood Khan, asked me for a pass for the screening. Many students of the Zoology Department also came to me, as they knew that I could use my 'contacts' in the Film Club to get them in. Anyway, their efforts were not really necessary, as the Hall was only moderately full, but when the film started, the auditorium reverberated with claps. Everyone enjoyed the classic that had won a record 11 Oscars in 1959. However, our ruse did not work for long as students began to flock to the auditorium, no matter the title of the movie.

Television had not yet arrived in our homes; it was the era of cinema halls. Aligarh had eight cinema halls in the 1970s and 1980s. In fact, our main entertainment was watching movies on the large screens of cinema halls, the most popular being Tasveer Mahal, situated close to the University. The Mahal would often show the so-called 'Muslim social' movies – the most popular being *Mere Mehboob* (1963), as it depicted AMU in the opening sequence. Even 10–15 years after its release, it would run to jam-packed houses for many days, despite the bedbugs biting the audience sitting on broken chairs! Contrary to its name, it was no fancy palace – Tasveer Mahal was the shabbiest among all the cinema halls of Aligarh. I remember one of my seniors in the Zoology Department doing research on the impact of different pesticides on bedbugs, and Tasveer Mahal and Aligarh Jail were the 'laboratories' of his experiments on these nasty little villains.

Soon after commencing my MSc in AMU, I realized that there were many hunters among the University staff, despite the fact that the Wildlife Protection Act, which had just been promulgated by the Indian Parliament in 1972, banned the trapping, selling and hunting of wild animals. My opposition to hunting, or rather poaching, by supposedly educated teachers was received with derision, with some claiming that it was their right and tradition, even religiously sanctioned activity, to hunt. While I could not do much during my

MSc period (mainly due to my study schedules), as soon as I joined the PhD programme I started my campaign more vigorously. It started as a one-man battle against a full army of hunter-teachers; gradually, some friends started supporting me. Prof. A.H. Musavi – a flamboyant teacher, famous for his fine taste in clothes, cigars and wine, and his eloquent talk, and who used to teach us evolution – also encouraged me. By this time, I was in regular touch with J.C. Daniel, curator of the BNHS, and Dr Sálim Ali on conservation issues. I wrote to them about the poaching cases in Aligarh, and both were prompt in replying to my letters – a habit that I learnt from them. They would encourage me to talk to the authorities to curtail poaching, which I started doing.

My conservation campaign took a dramatic turn when a teacher, who was also the warden of the V.M. Hall (Viqarul Mulk Hall), a residence close to Habib Hall, shot a peafowl that had drifted near his residence building. It was a summer afternoon, and he had been resting. But when he saw the peafowl, he was so excited that he came out of the house wearing his *pyjama* and *baniyan*, and shot the bird. Students who witnessed the crime caught him, and even wanted to beat him up for killing the national bird, but on realizing that he was their warden, they ran away, taking the dead peafowl along with them. Knowing my interest in conservation, a few of them came to my room, but by the time I went to the site, the drama was over. I got extremely upset and wrote a complaint to the Vice Chancellor about the 'criminality' of his staff member, naming the person. By the evening, the news spread across the entire campus, with some even claiming that I had written to the prime minister (PM)! The Vice Chancellor called the culprit and reprimanded him, further angering the hunter lobby. On learning that some of them wanted to physically assault me, I had to lie low for some days. But most people on campus agreed that the hunter had brought a bad name to the University by shooting the national bird.

This incident inspired me to establish a conservationist group or a club in the University. I did not get much support from the Zoology Department, where I was now doing my PhD, but the Drama, Literary and Music Club members came to my aid, and so did Dr Sarwar Rais of the Geology Department. We wanted a senior person to guide us, and Prof. Musavi was our first choice. He was initially reluctant to become the group's president, but we convinced him that no one was better suited; he accepted the offer munificently. I became the secretary of the Nature Conservation Society of Aligarh.

We started a quarterly newsletter in which we would write articles on local conservation issues, but we needed to conduct activities that would attract students and teachers to our cause. From my princely scholarship of Rs 500 a month (it had been increased from Rs 300 to Rs 500 after two years), I saved money to develop posters. The Art Club painted posters on wildlife and we soon started wildlife exhibitions in the Kennedy Hall. I also started subscribing to WWF-India, and from them we got greeting cards, calendars, and leopard and tiger cub posters to sell. The baby tiger poster, published in 1976 was a huge hit, followed by a baby leopard poster in 1978. In fact, all the posters, which we would sell at Rs 15 each, were very popular and hundreds of hostel rooms in the AMU campus came to be adorned with them. The sale proceeds would be sent to WWF, after retaining a small amount for our Nature Conservation Society.

I then found out that the Canadian High Commission in New Delhi could loan us wildlife documentaries, so we started showing these films once every two months. Prof. Saxena of the Botany Department was also a great supporter of our activities; he readily agreed to get us an old dusty projector from the Department's storeroom, and even opened a large lecture hall in the evenings for the viewing. Soon, we found a better projector in the Mathematics Department; its head, Prof. Ishar Hussain, was interested in nature, so he would bring his children and friends. What started as a lone fight for wildlife became a mini movement in the AMU campus within three to four years.

While all this was going on, my fight with poachers continued, as did their anger against me. During the Emergency (1975–1977), they plotted to put me behind bars under the notorious MISA or Maintenance of Internal Security Act, through the connivance of a corrupt police officer who was an accomplice in their poaching trips. A friend tipped me off about their secret plan, so for some months I stopped going out of the campus; if required, I would go with friends.

While the *Nature Conservation Society of Aligarh* was organizing lectures, film shows and exhibitions, I was in constant contact with the Chief Wildlife Warden (CWLW), V.B. Singh, regarding poaching in the Aligarh region. Whenever I came across any authentic news of shooting of Blackbuck or ducks, I would send in a complaint to the CWLW, sometimes copying the vice chancellor. At that time, there were no wildlife staff in the district, and the Aligarh-Agra region was looked after by a wildlife warden posted in Meerut. A kind gentleman, O.P. Tyagi, was a very effective warden and would often visit me in Aligarh. He also posted a wildlife guard in Aligarh who would visit most of the jheels regularly to prevent poaching as well as to demonstrate the presence of the forest department, a move which created fear among the poachers. Again, I was targeted as the 'culprit' who curtailed their nefarious activities. The Vice Chancellor of AMU, Prof. A.M. Khusro, a distinguished economist and a genial man, came to our exhibitions many times, and it was his support to our cause that sent a message to the criminal poachers that it would not be easy to jeopardize a conservation mission started by students and research scholars. The campus poachers then adopted a new tactic – besmirching my name on a vacuous plea, proclaiming that Asad Rahmani was spoiling AMU's name by writing letters to authorities and by writing about their activities in the newspapers.

One day, while cycling to Habib Hall I saw that the large Sheesham trees lining the road to the impressive V.M. Hall were being cut. Enquiries revealed that the new wildlife warden wanted to cut the trees to 'beautify the entrance of the V.M. Hall, as the trees were old'. A protest letter to the Vice Chancellor could not stop the wanton destruction of the old trees, but it resulted in some policy changes in the University Garden Department. The Property Officer, Mr Khalique Ahmad Siddique, met our group and promised that in the future no tree would be cut for such silly reasons. When this news spread in the campus, our tree protection campaign resulted in a huge goodwill towards our *Nature Conservation Society*. Many students came forward to join our Society or came to our functions. Another game changer was visits of

Dr Sálim Ali to Aligarh. Aligarh Muslim University was the first university to recognize his contribution to ornithology by awarding him an honorary PhD in 1958.

Activities of our *Nature Conservation Society* sowed the seed of much bigger things to come, which finally resulted in the establishment of the Department of Wildlife Sciences. It would not be out of place to tell the whole story here, one that could well be called *The Sacrifice of a Peafowl for a Noble Cause*.

―

The seed that I sowed in AMU in the mid-1970s, by establishing the *Nature Conservation Society of Aligarh*, was nourished by Prof. Musavi. His ambition was to start a Department of Wildlife, but he faced strong opposition from the Zoology Department. But, Prof. Musavi was able to convince the Academic Council to offer a paper on wildlife sciences. Later, under the sixth five-year plan, a scheme was sanctioned for the Zoology Department to start a specialized course on wildlife, and if the course was successful, the University Grant Commission (UGC) would agree to support it further. The course was known as 'Introduction of Wildlife Biology and Ornithology in Indian Universities'. Prof. Musavi worked tirelessly, and single-handedly, to make it successful, which it was, as later developments show.

In 1986, Prof. Musavi organized an international seminar: 'The Role of Universities in Wildlife Education and Research'. It was attended by Dr Sálim Ali, J.C. Daniel, David Ferguson of the US Fish and Wildlife Service (USFWS), H.S. Panwar, the first director of the newly established Wildlife Institute of

India (WII), along with V. Savarkar, Dr A.J.T. Johnsingh, Dr Allan Rodgers and Dr John Sale from the WII faculty. Dr V.S. Vijayan, S.A. Hussain and I represented BNHS. The overt purpose was to discuss emerging issues related to wildlife, but the hidden agenda was to convince the University to create a new department – the Department of Wildlife Sciences. On the second day, in the concluding discussion, Prof. Musavi asked Dr Sálim Ali, about his opinion on this matter. Dr Ali, being a life-long supporter of new initiatives, suggested that the Vice Chancellor (who was present at the seminar) should start the new department. The deal was clinched with applause from the audience. The Vice Chancellor, too, became a strong supporter of the new initiative after meeting Dr Ali, H.S. Panwar, David Ferguson, J.C. Daniel and others.

So, despite some opposition from the Zoology Department, all things were converging in one direction – the establishment of a new department in AMU! The great University, famous for teaching traditional and modern subjects, finally had the impetus to start a full-fledged course in wildlife. As Victor Hugo, French poet and novelist, said, 'No force on earth can stop an idea whose time has come.'

In the last 35 years, what is now the Department of Wildlife Sciences has produced many good wildlife researchers, leaders, forest officers and teachers. In November 2023, visiting the impressively renovated building of the Department, I was pleasantly surprised to see a flock of five peahens and one young peacock calmly foraging in the lawn, some 10–12 m from the students. I remembered the sacrifice of a poor *morni* 45 years ago, who had made the fatal mistake of appearing on the boundary wall of V.M. Hall, only to be greeted with a hail of bullets. Should I thank her for electrifying the conservation movement of AMU? Her sacrifice has not gone in vain.

―

In 1977, AMU started a one-year diploma in journalism. Interested as I was in reading and writing, I signed up for it, as the classes were held in the evenings. My PhD supervisor, who was already angry with me for 'wasting time' in conservation activities, was exasperated when he came to know that I was attending evening classes for a journalism diploma as well. We were seven students in that first batch, and I topped the class. During the course, we invited a few luminaries, such as veteran journalist and commentator Kuldip Nayar and the famous writer and Urdu essayist Ismat Chughtai, to address the students. I remember visiting Ismat Chughtai to invite her to deliver a lecture to the diploma students. Despite being a celebrated writer, and extremely busy, she agreed. I was a great fan of her work, as she was a rebel representing suppressed Muslim women, and wrote extensively on female sexuality, middle-class gentility and exploitation of the working class. Part of the Progressive Writers' Association, which had some of the finest Urdu and Hindi writers and poets, she started writing bold short stories in the 1930s, when most Muslim women were in *purdah*.

For love of the written word, the first thing that I bought from my own earnings was a typewriter. In 1974, from my monthly scholarship of Rs 300, I saved money for a few months and bought an old Remington Portable. Before this acquisition, I used to go to the Aligarh court to get my articles typed. My first published piece, a small letter, 'Our Lions' was published in the *Pioneer* newspaper on 28 October 1971. (At that time, we were in Lucknow, as my father, after retiring as a district judge, was posted there as the Chairman of the Sunni Wakf Board.) I was thrilled and read it many times, but no one in the family noticed the article. In the evening, I proudly showed it to my eldest brother, Khalid, who had come from Bombay, where he worked for Union Bank. He congratulated me, and laconically advised me to write more often. That's it. I presume most putative writers have similar beginnings.

In addition to the Lucknow *Pioneer*, I sent my typewritten articles to newspapers, such as the *National Herald*, *Indian Express* and *Hindustan Times* (all in New Delhi), and the *Science Reporter* magazine (also based in New Delhi). Sometimes I would get Rs 50 or 60 for an article, a princely sum at that time. The highest amount that I got was Rs 120; it was from the *Indian Express* for an article on the Kiang or Tibetan Wild Ass in 1979. From this additional income, in 1975, I bought a second-hand Zenith camera from my friend Saeed Ahmad, who was fond of photography and had his own darkroom for developing black-and-white pictures. We borrowed a 300 mm lens from Prof. Hamida Saiduzzafar, the head of the Department of Ophthalmology, AMU (who was also interested in wildlife), and went to Bharatpur to photograph birds. In the mid-1970s, owning a camera and a long lens meant that you were a rich man! To us, it did not matter that the camera was an old Russian brand, and the lens was a borrowed one! We both strutted around in the sanctuary with our new-found status symbols.

Between 1972 and 1980, I published nearly 100 articles and letters in newspapers and magazines. Many articles were based on my personal field experiences (whatever I could manage at that time) and many others were based on extensive reading. The Maulana Azad Library at AMU used to get magazines such as *Science, New Scientist, Nature, Reader's Digest, Time Magazine, National Geographic* and so on. I would spend at least two hours in the evening, twice or thrice a week, filling my notebooks. I still have those notebooks, and often read the notes made 50 years ago. They still enlighten me – such is the power of the written word!

4

The Early Conservation Struggles

During my MSc days at the hostel in AMU, when I could not afford to subscribe to a newspaper, I would visit the hostel's common room daily to read at least two newspapers. Later, with the Department of Science and Technology's scholarship money in hand, I started subscribing to the *Times of India*, and would also purchase second-hand magazines. In the 1970s, the conservation movement was developing, so news on environment and wildlife started appearing in the popular newspapers and magazines. This inspired my own early conservation crusades. I would read any wildlife article published in available newspapers. Besides M. Krishnan, who used to write in *The Statesman*, my other favourite was U.C. Chopra, who wrote in the *National Herald*.

I can claim that I was the first person to write about the danger posed by the Mathura Refinery to the Taj Mahal and other nearby monuments due to air pollution. Without claiming to be an authority on pollution, my May 1976 article in the now defunct *Imprint* magazine[2] used basic science to argue that if the refinery released sulphuric acid (H_2SO_4), it would have a corrosive chemical action on marble, which is basically made of calcite ($CaCO_3$). *Imprint* ran my piece with a dramatic title 'Should we kill the Siberian Crane? Should we destroy the Taj Mahal?'

On 12 February 1978, my article 'Threat to Taj Mahal' was published by *Patriot* magazine[3] and on 1 March 1979, another article 'Save the Taj' was published in a little-known magazine called *Concept*.[4] My articles drew the attention of Prof. T. Shivaji Rao, noted environmentalist and author of the book *Save Taj Mahal and People of Braj Mandal*. Prof. Rao (who later became the honorary director of the Centre for Environmental Studies, GITAM [Deemed to be University], Visakhapatnam) was one of the first environmentalists to write about the dangers of pollution, and was also an early crusader of human rights. He, like me, was worried about the location of the oil refinery within the zone of influence of the Taj Trapezium Zone.

Around this time, a symposium was organized at Agra on the danger to the Taj Mahal and the mitigatory measures required to protect this famous monument. Prof. Rao invited me for the symposium, and I attended, though I was a bit star-struck meeting the high-profile delegates within the precincts of a five-star hotel (my first time in one!). Prof. Rao welcomed me graciously and told the delegates about me. A pat on the back by an expert like Prof. Rao told me that I was on the right track with my writings and initiatives. Thanks to his efforts, proper mitigatory measures were adopted at the design stage of the refinery, so fortunately no harm was done to the Taj Mahal. Incidentally, in 1998, about 20 years after the commissioning of the Mathura Refinery, I did a study on birds, which resulted in a small booklet and later a coffee-table book, *Birds of Mathura Refinery*.[5] In an ironical twist to the environmental concerns raised earlier, a small breeding colony of Painted Storks, Asian Spoonbills, and various species of egrets and cormorants has now come up on the Acacia trees growing in the refinery's large tanks, where water is cooled. Officials of the Mathura Refinery proudly show this colony as their contribution to wildlife conservation!

Another campaign that I undertook around that time was for the protection of waterfowl near Narora in Bulandshahr district, about 60 km from Aligarh. I had been going to Narora, a small town located on the bank of River Ganga, since 1975. My destination was not the town but the river; a long barrage across it had resulted in a large reservoir (nearly 2 km long, upstream), which used to attract a large number of waterfowl. The bus that plied between Aligarh and Moradabad would go via Narora (the barrage has a public road) so I would take a bus to Narora; after getting down at there, I would walk along a bund for a kilometre, followed by a 10–12 km long trek on a kachcha road on the other bank of the reservoir. This peaceful trail beside the mighty Ganga was my favourite birding area in the 1970s. Sugarcane, wheat and mustard fields, large *jheels* of spill-over water, secluded tree groves and grazing lands provided an ideal rural setting for a winter outing. The icing on the cake was thousands of Bar-headed and Greylag Geese, Northern Pintails, Northern Shovelers, Eurasian Wigeon, Gadwall, Common, Tufted, Red-crested and White-eyed Pochards, Common Teal, occasional flocks of Cotton Pygmy Geese, Common and Demoiselle Cranes, and many species of terns, gulls and raptors. I would make notes of all the birds seen during my visit.

Narora was where I saw birds hitting high tension wires for the first time. Another time, when I went there with my photographer friend, Saeed Ahmad, we found an otter trapped in a fishing net. We cut the net and released the poor animal. I wrote about the birds of Narora to J.C. Daniel and Dr Sálim Ali and got in touch with the regional director of UP Tourism, based in NOIDA, proposing that the Narora reservoir be declared a bird sanctuary. The young Regional Director, named R. Prasad, was interested and even visited the area with me in 1978. A derelict guest house beside the Ganga bank was renovated for tourists, and a large garden built in front of it. The best part was being able to take Dr Sálim Ali to Narora, in December 1978. In his characteristic style, he said, 'It is a natural sanctuary and should be protected.' A photograph of Dr Sálim Ali watching birds in Narora was published alongside an article I wrote for the *Tourism and Wildlife* journal, called 'Narora – An Ideal Place for Bird Watchers'.[6]

However, the large congregations of birds also attracted hunters. One of them, based in Aligarh, taunted me by sharing that he knew the engineer of the Irrigation Department, who gave him official boats to hunt the birds. I wrote to the Department about this. That missive and my personal requests to the engineers to not provide boats to hunters stopped the poaching to a large extent. Media outreach also helped – 'Poachers of Narora' appeared in the *National Herald* on 6 June 1977.[7] I sent the cutting to the CWLW of UP, requesting him to protect the birds of Narora reservoir. Back then, photocopying services were not available in Aligarh, and I would buy three copies of the newspapers that carried my articles in order to send the clippings along with my letters. I have always believed that if we want conservation action on the ground, we have to reach out to the relevant authorities. O.P. Tyagi, the regional warden, also came to Narora to meet the engineers. Almost a complete stop to poaching took place soon afterwards.

After collecting bird data for four years, I wrote my first long paper in the *JBNHS* (1981)[8] – 'Narora Reservoir, U.P.: A Potential Bird Sanctuary'. Another paper 'Narora Reservoir: An Excellent Habitat for Waterfowl' followed in the *Corsonat*[9] in 1989.

In the early 1970s, conservationists became worried about the impact of radiation on animals and people from a proposed atomic power plant near Narora. However, on 12 March 1989, 35 years after the power plant went critical (in the operation of a nuclear reactor, criticality is the state in which a nuclear chain reaction is self-sustaining – that is, when reactivity is zero), its operation has been incident-free. In fact, the creation of a large buffer zone around the power plant (for security reasons) and the planting of thousands of trees has resulted in a 'mini sanctuary', where Cheetal (*Axis axis*), Sambar (*Rusa unicolor*), Nilgai (*Boselaphus tragocamelus*), Wild Boar (*Sus scrofa*), Golden Jackal (*Canis aureus*), Rufous-necked Hare (*Lepus nigricollis*) and many species of birds find safety. The release of warm water from the plant into the mid-Ganga canal also resulted in the development of huge Muggar or Marsh Crocodile (*Crocodilus palustris*). As Muggars, like other reptiles, are cold-blooded animals (poikilotherms in technical terms), they get body heat from sunrays. I was told that if they live in warm waters, they grow fast and become rather large. A wetland, called the Hakimpur Jheel (also in the buffer zone), has become a bird sanctuary and a showpiece of environmental conservation for the authorities of the atomic power plant. The wetland is named after a relocated village and falls in the exclusion zone (where no outsider is allowed) of the nuclear power plant, and hence is very well protected.

My junior at AMU, A.I. Siddique, was the public relations officer of the Nuclear Power Corporation of India Ltd (NPCIL). In early 2005, he visited me in BNHS. We went on to develop the Environmental Stewardship Programme (ESP), under which BNHS and NPCIL conducted environmental education workshops in seven atomic power stations across India. Under this programme, I visited Narora that same year, when Siddique introduced me to two young men, Jitendra Pandey and Raja Mandal. Full of energy and vigour for the protection of the environment, these young men, for whom I coined the moniker Mandal–Pandey (after the hero of the 1857 war of Independence), turned out to be as inspirational as their close namesake.

Both work at the Narora Atomic Power Plant and during their spare time, they work for wildlife conservation at the grassroot level, be it surveys of the Indian Skimmer and Gangetic Dolphin (*Platanista gangeticus*), or organizing student rallies against plastic pollution. Through their dedication and hard work, they have convinced their bosses in the atomic power plant that the exclusion zone can be successfully converted into a wildlife sanctuary. The result is that Narora now has a thriving wetland full of resident and migratory birds, canals full of crocodiles and turtles, and forests full of Nilgai, Black-naped Hares and Wild Boars. There are unconfirmed reports of Swamp Deer being spotted in the vast dense grassland, which has come up in the low-lying areas near the nuclear plant. Mandal and Pandey keep track of the nesting success of the Black-necked Stork (*Ephippiorhynchus asiaticus*) and the movements of the Gangetic Dolphin. They also chase turtle poachers, give lectures in local schools on conservation, talk with housewives to minimize the use of non-degradable items, keep records of birds, take socio-economic data of nearby villages, study changes in agriculture practices and help visiting conservationists. After initial scepticism, their bosses, all engineers and nuclear scientists, now greatly appreciate their conservation efforts. Neither Mandal nor Pandey had a biology background, and before attending the BNHS workshops, neither knew much about wildlife conservation, underscoring the value of such workshops and lectures. May we have more people like Mandal and Pandey to fight conservation battles at the grassroot level!

Thanks to the attention given to the Narora reservoir and its surrounding areas, and the presence of a good population of Gangetic Dolphins near Karanwas village, at the initiative of WWF-India, an area of 266 sq. km. – from Brijghat, Garh Mukteswar, to Narora in Bulandshahr (a linear distance of 85 km) – was declared as a Ramsar Site in 2005. I cannot take credit for this but my early visits to Narora in the mid-1970s and my early fights to keep it free of poachers could have played some role in preserving the wildlife of this place.

One credit that I am proud to take is for the protection of the Sheikha Jheel. From being overrun by criminal poachers in the early 1970s to its declaration as a wildlife sanctuary in 2016, my frequent letters and articles played a key role in its conservation.

Located 17 km from Aligarh on the Aligarh–Jalali road near Sheikha and Bhawan Khera villages, the original Jheel was divided into three parts when the Upper Ganga Canal was constructed in 1852. Sheikha Jheel is a typical monsoonal wetland of the Gangetic plain. It gets most of its water from rainfall, but seepage of water from the adjoining canal has made it perennial. I first heard of the Sheikha Jheel in 1974–75, when I was doing my PhD in AMU. In 1975, I went there in a rickety overcrowded bus. The area was notorious for looting and dacoity, and local villagers warned me to return before dark. Despite the morbid warnings and threats from hunters (who had a score to settle with me, as I was trying to stop their nefarious activities), my introduction to Sheikha Jheel started on a memorable note – I saw a pair of Sarus with a juvenile, and thousands of ducks.

A few letters to the CWLW of UP to protect Sheikha Jheel resulted in the visit of the Wildlife Warden of the Agra–Meerut region, O.P. Tyagi, who agreed to post a guard at Sheikha. This brought some control on rampant poaching. A meeting with villagers ended with assurances that they would not allow hunting anymore. They even started chasing out hunters, who (rightly) blamed the University's young researcher for spoiling their 'sport'. In the mid-1980s, after the Centre of Wildlife was established in AMU, Sheikha Jheel became like an open-air classroom to teach bird identification to students. In the 1990s, it was included in the list of Wetlands of National Importance (under the Ministry of Environment, Forests and Climate Change [MoEFCC]), and in 2016 it was declared as a bird sanctuary. My habit of writing detailed field notes in my diary revealed an interesting element. The page for 5 March 2006 says,

Delighted to visit Sheikha Jheel. My dream to see this wonderful wetland protected has come true after 30 years! I heard about this Jheel in 1975–76 and since then I have been trying to protect it. Wonderful to see the signboard of Sheikha Jheel, a tall (and dangerous) *machan*, a forest guard, a nature trail and more than 6,000 birds peacefully enjoying the facilities.

Nearly 200 species of birds, both aquatic and terrestrial, are found in and around the Sheikha Jheel. During winter months, sometimes more than 5,000–6,000 birds are found and, in a year, probably 20,000 birds use this wetland, with large migratory flocks of waders seen in March–April. For many species, it is an important stopover site. During my second stint in AMU from 1991 to mid 1997, a local conservationist, Subodh Nandan Sharma, and I wrote a management plan to integrate and enhance the natural habitat of Sheikha Jheel. While not all our recommendations were accepted, I am happy that better protection was provided, with regular guards being employed to patrol the area 24x7. Now, Sheikha Jheel looks like a typical waterfowl sanctuary, with thousands of birds and a large number of visitors, though most visitors come to enjoy the scenery and the shaded pathways, and only a few visit for birdwatching. In the nature-deprived Aligarh district, Sheikha Jheel shines. Even a hotel has come up across the road, indicating the popularity of this once-neglected wetland.

I am afraid Sheikha's popularity could become its bane too. Firstly, during a recent visit in 2023, I heard that the district administration wants to 'develop' Sheikha Jheel by including some adjoining areas and planting trees, and developing new infrastructure. People forget that it is a wetland and should remain as a wetland. Secondly, it is a bird sanctuary, not a picnic spot, so the first priority should be birds, not tourists, and its pellucid water should be protected from unwanted developments. The funds available with the administration should instead be used for the total removal of the pernicious 'cancer', which botanists call water hyacinth. This invasive plant from South America is a bane of our wetlands, as it grows profusely and covers large areas quickly, choking the water. It had totally covered Sheikha Jheel in 2022, but fortunately the Forest Department took measures and cleared almost half of the Jheel, before funds ran out. There should be a regular programme for its removal, and it should be removed, particularly during the summer and early monsoons, when migratory birds are absent.

Development around the Jheel also needs to be monitored – Sheikha Jheel is only a 25 ha area, and is surrounded by villages, canals and agricultural fields. Its large catchment area needs to be protected, and any 'development' in and around the Jheel should be closely monitored. Nearly 40 years ago, I was instrumental in saving Sheikha Jheel from hunters. Now, we must save it from the *vikas* bandwagon!

5

Joining BNHS

The advertisement seeking applicants for the post of a researcher at BNHS appeared in newspapers on the day of a solar eclipse in 1980. A large number of candidates were called for an interview at the venerable Hornbill House, the headquarters of BNHS, where a formidable interview panel consisting of Dr Sálim Ali, J.C. Daniel, Prof. C.V. Kulkarni, S.A. Hussain, A.N.D. Nanavati, Prof. P.V. Bole, among others, awaited them. This panel could have intimidated anyone, but my interview went very well, since some of them knew of my keen interest in wildlife. In a few weeks, an appointment letter from BNHS appeared in the postbox, and I was summoned to Point Calimere Sanctuary in Tamil Nadu, where our field training was planned. I suspect the selected candidates were called to that remote sanctuary to test their endurance.

The two-day trip by train from Lucknow to Chennai – then an overnight train journey to Thiruthuraipoondi, followed by a bus ride to the sanctuary, 42 km away – left me exhausted. Plus, I was weighed down by my old Remington typewriter, a heavy Russian camera, an equally heavy pair of binoculars and books, in addition to clothes. A warm welcome from P.B. Shekar of BNHS, deputed by J.C. Daniel to make us comfortable, dissipated all the tiredness. We were a mixed group of young researchers, four with PhDs, six or seven with newly acquired MSc degrees and seven or eight with BSc degrees. We were housed in a large building called Subburaman Illam. Dr Sreedhar, a Tamil brahmin and a fellow researcher, who was my age, became a good friend, and inducted me into Tamil customs and cuisine during that Calimere stint.

At Calimere, we were taught bird ringing by P.B. Shekar, who had been doing it for two decades. Then there was R. Sugathan and S.A. Hussain, also of the BNHS, who came to Point Calimere to teach us bird ringing; Hussain was an old hand on the subject. The first bird I ever ringed was an Oriental Magpie Robin (*Copsychus saularis*). We were ringing birds under the Bird Migration Project, funded by the USFWS, and were supervised by Hussain, though Dr Sálim Ali was the principal investigator. We also learnt how to carry out bird and vegetation censuses, as part of a rigorous training programme that ran from pre-dawn to post-dusk in the forests and salt pans of Point Calimere. The sanctuary had a population of Blackbuck – the animal on which I wanted to do my PhD – and on Sundays, after washing clothes and writing notes, I would go alone to the sanctuary to watch them. Another pastime was to go to the beach to collect shells; for a person from North India, the sea coast was a great draw.

Also, as a North Indian, I would wear a *kurta-pyjama* during non-office hours. Once on a hot Sunday afternoon, I heard a racket outside Subburaman Illam – it turned out that a pre-fledged crow had fallen from its nest, causing its parents and their companions to create a commotion. I picked up the injured young bird, intending to place it on a high branch, away from the predatory cats that roamed the neighbourhood. As I was trying to move the young bird, the crows started cawing even more, and worse, they began to dive-bomb me, no doubt reading mala fide motives in my altruistic act. This drew the attention of the people in the sleepy village. Since we were known as *paravai-aarvalar* (bird lovers), some youngsters came to my rescue. From then on, whenever I stepped out of the house, the cawing and dive-bombing would start. Even when I went to the village, the crows would express their anger vocally – clearly the crow telegraph had flashed the intruder alert most effectively. For the entire period of my stay in Point Calimere, I had to face this ignominy. But it also taught me about the intelligence of crows; because even when I was dressed in a different attire – not the conspicuous white *kurta-pyjama* – the crows would inflict their annoyance on me. It clearly proved that they could recognize the face, no matter what the disguise! Twenty years later, I returned to Point Calimere as BNHS director – by then, fortunately, a new generation of crows had come up, and I could peacefully walk around the village, reliving my memories, including those that involved being insulted by their ancestors!

Bombay Natural History Society has a long association with Point Calimere Sanctuary, and its team began ringing birds there as far back as 1969. In the 1980s, BNHS established a field office in Thambuswamy Illam, an old forest rest house, where Dr Sálim Ali stayed when he visited Calimere. The Sanctuary is a field laboratory for BNHS and local colleges, particularly the Anbanathapuram Vahayara Charities (AVC) College of Arts and Sciences, Mayiladuthurai, which started the first MSc wildlife course in India. The Sanctuary has also produced many good scientists, but the finest scientist of the crop is undoubtedly Dr S. Balachandran of BNHS, who has spent most of his career there. He was my colleague for more than 40 years. In 2007, he established a permanent ringing station in Point Calimere, raising funds single-handedly.

But fieldwork is not for everyone. Some of the trainees left within four to five weeks, but more joined us, and continued their journey with BNHS. One of the most impressive additions to our group was a 21-year-old young man who came with a WWF rucksack slung on his back. Ajai Saxena was impressive – having completed his BSc from Allahabad University, he wanted to make a career in wildlife. Ajai picked up wildlife quickly – a trait not seen in some others who also joined BNHS for a job! After three months, a few of us, including Ajai and me, were shifted to Bharatpur, Rajasthan, where we lived in Lal Kothi, given to BNHS by the erstwhile Maharaja. Ajai left BNHS in 1981 to appear for the Indian Forest Service (IFS) exams; as expected, he went on to carve out a successful career in the IFS and retired as the principal chief conservator of forests, Goa, a few years ago.

At Bharatpur's Keoladeo National Park (KNP), a sanctuary at that time, I worked on the Bird Migration Project for a little more than a year, after which I was shifted to Endangered Species Project. The main focus of the latter was the Great Indian Bustard and Asian Elephant. It was a project that would change my life.

6

The Bird That Changed My Life

In 1980, Harsh Vardhan, the president of Tourism and Wildlife Society of India (based in Jaipur), organized an international conference on the Great Indian Bustard (GIB) in the aftermath of an Arab sheikh's attempt to hunt the Houbara Bustard in the Thar Desert, where the GIB also lives. The conference publicized the then status of the GIB in various parts of India, including the discovery of a few in Solapur, Maharashtra, where in 1979, after about 100 ha of barren, overgrazed land (3 km from Nannaj village) was taken over by the Forest Department and grazing stopped, a large bird was seen, which no one could identify. B.S. Kulkarni, the principal of a local school in Solapur, who had an interest in birds, was contacted. He identified it as the 'Maldhok', which is the local name for the GIB. Soon after, the Forest Department conducted a survey to identify the Maldhok areas in Maharashtra; the results of this were presented during the 1980 conference in Jaipur.[10]

In 1981 began a series of surveys, as part of the Endangered Species Project of BNHS. In April of that year, I went to Solapur by train and from there I was taken to Nannaj, about

20 km away, where a few GIB were usually seen during the monsoon in the grassland plots developed under the government's Drought-Prone Areas Programme (DPAP). The Range Forest Officer (RFO) took me around in his jeep. We did not see any bustards but a few Blackbuck, one of my favourite animals, were around.

The next destination, in the month of May 1981, was Sonkhaliya, around 10 km from Nazirabad in Ajmer district. With requisite permission from the Forest Department in hand, our small team (which included my assistant Jugal Kishore Gajja and our driver Mohanan) met the Ajmer RFO who advised us to meet a certain Ranvir Singh Rathore in Bandanvara (in Ajmer district). He turned out to be the most knowledgeable person as far as bustards of that area were concerned. This is what I wrote in my field notebook that day:

4.30 p.m.: Went to Sonkhaliya with Mr Rathore. After few miles of metallic road, a *kachcha* road started, so Mr Rathore drove the jeep, because he knew the terrain. From Sonkhaliya [we] picked up Goga, a driver and a 'local authority' on bustards. Mr Rathore and Goga were very sure of seeing bustards. Roamed for two hours around the best habitat of the GIB but in vain. The next morning, we started early, at 5.55 a.m., along with Mr Rathore, and saw the first bustard of my life, a juvenile that took off and settled 1.50 km away. Soon we saw four bustards – one male and three females. In the next two hours, we saw a total of 15 bustards.

It was my first sighting of the GIB! Mr Rathore was careful not to disturb the birds, so we maintained a good distance, enjoying the grand sight with binoculars. We dropped Goga off at his village, where his 7-year-old son, Ganesh, came running to greet his father. Thanks to the publicity around the GIB, Goga was a celebrity in the area, as he was much in demand by the Forest Department to accompany dignitaries who wished to see the bustards. I must say, Mr Rathore had trained Goga very well. Tall, dark and handsome, with a red–maroon *paggar* (traditional headgear), long white *kurta* and white *dhoti*, *mojari* (traditional footwear) and a curling moustache, Goga made an impressive figure. Like the bustard, Goga left a lifelong impression on me.

It was early July 1981 when I reached Karera, a sleepy little tehsil town, around 45 km from Jhansi, on the Shivpuri–Jhansi road. The discovery of bustards in the vicinity had given it some publicity; its other important landmark was an old derelict fort perched on a hill. It was said to have been built by the Paramaras, who ruled the area under the aegis of the Mughal empire. The fort had passed through several hands, including the ruling family of Jhansi; despite its colourful history, the fort was neglected and in ruins. If I remember correctly, I managed to see three bustards in the Karera Bustard Sanctuary, which was established to protect them.

After Karera, I returned to Bombay. But soon, in the company of two of my assistants, Ranjit Manakadan and Jugal Kishor Gajja, I went to Jaisalmer (in end July 1981). Unfortunately, we returned without seeing a single bustard, due to the non-availability of a suitable vehicle to go to the remote areas. So, in August, along with the assistants, I returned to Nannaj for fieldwork. A total of eight bustards, including two displaying males, were seen in the Nannaj grasslands. I decided to study bustards at Nannaj because they were relatively easily visible. First, we searched for accommodation in Nannaj. But being a small village, rooms were not available on rent; for three months, the three of us stayed in a cheap hotel in Solapur and would commute to Nannaj by bus every day, sometimes making two trips a day. As staying in a hotel, even a cheap one, was proving expensive and difficult in the long run, we finally managed to hire a house. For five months, we spent long hours watching the bustards, and this gave us a good preliminary data.

The study continued for four years, with Ranjit being posted there, and Jugal leaving the project a year on. We would spend almost the whole day in the field. Initially, the bustards would hide from us and were quite difficult to sight, but after a few weeks the birds started tolerating us. A male bustard has a large territory that he defends from the other adult males. We witnessed two adult male territories, one inside the protected grasslands and one outside. The one inside was relatively undisturbed, so the adult male, whom we named Alpha, would display for many hours in the mornings and evenings. Females and young males were tolerated in the territory. During the first year of our study, we could only see the display from afar, but in subsequent years we managed to observe them properly at closer quarters, which helped us describe their elaborate displays in our reports and papers. We also saw the mating of bustards, a seldom observed behaviour that, until then, had not been described properly in published literature. It was the golden period of my bustard study. Some of the finest nesting, chick survival and behaviour data were taken at Nannaj, which we eventually published in a series of research papers and popular articles.[11-16]

The next year (1982) I visited Sonkhaliya again, with the

intention of establishing a second field station, but we were denied permission by the then CWLW Mr Kailash Sankhala on the plea that 'you will disturb the birds'. I met Rathore and Goga, and enquired about Ganesh. Goga replied matter-of-factly that Ganesh had died a few months previously, after getting an infection. Life was too short for the little boy. Here was Goga trying hard to save the bustard, but the poor man could not save his own son due to a lack of good medical facilities. At that moment, I realized the reality of poor Indians. Not much has changed in 40 years.

Harsh Vardhan told me of some bustards surviving in the Kota region, and I was given the contact details of one Bharat Singh of Kundanpur, who could help us. By 1981, BNHS had acquired a new vehicle for its newly launched Bird Hazard Project in Agra. Since the vehicle had to be transported from Bombay to Agra, I suggested to J.C. Daniel that I could use this opportunity to visit Kota, Shivpuri and Gwalior districts en route, where bustards were spotted. He readily agreed. Goutam Narayan, who was a part of the Bird Hazard Project, and I set out from Bombay on 15 December 1981, reaching Kundanpur village, Kota, on 17 December. We were told that Bharat Singh was the *sarpanch*, so we were expecting a villager clad in *kurta* and *dhoti*, and maybe a traditional headgear. The person we met was a smart, highly educated man, who spoke fluent English. A product of the famous Mayo College, Ajmer, Bharat Singh was a very refined and cultured man, and was extremely interested in wildlife conservation. As it was quite late in the evening, he arranged our stay in his ancestral house in Kundanpur village. Over discussions that evening, he told us that during the previous 12 months, around 12 bustards had been seen in Sorsan *buld* (a flat stony plateau), just across the stream as the crow flies (but 12 km by road). He had kept detailed notes that are still a valuable part of my archival field notes. The next day we drove to Sorsan *buld*, but could not see any bustard. Although Blackbuck and Chinkara (*Gazella bennetti*) were in plenty, thanks to the protection provided by Bharat Singh.

From Sorsan, we drove via Shivpuri to reach Karera, where I had seen three bustards in July 1981, between Fatehpur and Ronija villages. It was difficult to spot the birds in the undulating landscape, which was also heavily overgrazed and punctuated with crop fields wherever a well could be dug up. People were extremely poor, with no electricity, dispensaries, schools or running water. And for most of the year, it was a water-starved area. However, I found it a good study site. The lone forest guard told me that there were 10–12 bustards, seen mainly during the monsoon. A dozen Blackbuck and two to three Chinkara enlivened the barren landscape. I decided to establish a field station at Karera, from where the bustard area was 20 km away (between Fatehpur and Ronija villages). After spending two days in Karera, we drove to Bharatpur, stopping at the Ghatigaon Bustard Sanctuary where five to seven bustards had been reported – we saw none on that December 1981 trip, but on subsequent visits I recall seeing six birds in the Kalitallia area, and a few others in the Tigra area near Gwalior.

In the winter of 1981–82, Ranjit, Jugal and I travelled to the Thar Desert to conduct a bustard survey. A major handicap this time as well was not having our own vehicle; we therefore stayed in a *tehsil* town called Phalodi for three weeks, and would travel by bus to the nearby areas where bustards had been reported, such as Bap, Nachna, Nokh, Nokhra and Khara. After reaching these villages, we would meet the villagers, and as directed by them we would walk down to the areas where bustards were sometimes seen by locals. Not all villagers were helpful; hearing my name, some would look quite suspiciously at me! I soon understood the reason for this. A few years previously, an Arab sheikh had come to hunt the Houbara bustard or *Tiloor* (in the local language). But the media reported this as 'bustard shooting', and for most people the word bustard meant the GIB! Anyway, it was impossible to survey the vast desert without a vehicle. The same problem plagued us when we attempted to survey the 3,162 sq. km Desert National Park, located in Jaisalmer and Barmer districts, to study the bustards, often in remote areas. We reported to J.C. Daniel and Dr Sálim Ali that, without a vehicle, it would be difficult to work on the bustard, as it survives only in remote areas. J.C. Daniel understood our concern and soon provided us an old vehicle that was lying unused in BNHS. Quickly repaired and made motorable, the jalopy – whom I christened the 'Old Girl' – became the companion of many an adventure.

'Old Girl' was put to the test when, on a hot May day in 1982, I set out to establish a second field station in the Karera Bustard

Sanctuary, accompanied by a young BSc researcher named Bharat Bhushan who had been newly recruited by BNHS. In those days, it took three days to cover the distance of 1,200 km from Bombay to Karera. J.C. Daniel had advised the mechanic (who repaired Old Girl) to fix the speed controller, so that the vehicle would not go beyond 50 km/hour. I thought it was unnecessary, as the old vehicle would only move at the gentle pace of 40 km/hour! On top of it, the roads were narrow and full of trucks, with drivers under the impression that the road was built only for them.

On 24 May, when we reached Karera in the mid-afternoon and the ranger arranged for our stay in the government guest house. Early the next morning, we started our bustard survey in the Sanctuary, a project that would continue for the next five years. Initially, the villagers of Fatehpur thought that we had come to see the bustard and would go back after the visit, but when they saw us visiting over the course of several days, they were surprised to know that a *daktar* (doctor) from Bombay had come to 'work' on the *Sonchidiya* (local name of GIB), a work that would go on for many years. Some thought I was crazy to leave the glamorous world of Bombay and live with them in a hot, small, dusty and poor village.

Fatehpur was 20 km from Karera town, reached only by a narrow dirt track, which barely resembled a 'road'. As diesel was not available in Karera at that time, I had to conserve fuel. So, even after completing the morning's fieldwork, I would remain in the field so that I could study bustards in the evening, sheltering under trees during hot midday when temperatures would go up to 48°C. I had never faced such deprivation before (the village lacked even basic necessities, like clean water), but I decided that there was no going back.

My leftist and socialist leaning gave me strength. I thought that when these poor villagers could live without electricity, running water and proper shade and food, why couldn't I?

But we still needed a house in Karera where we could keep our household items, books and files, and also cook, take a bath and get letters/communications by post. A kind pandit gave us his newly constructed house, which he had named Bapu Sadan, located beside the main Shivpuri–Jhansi road. The only condition was that non-vegetarian food should not be cooked in his house. The edict was easy for me to follow, as I was largely a vegetarian. From Karera, I would go to Fatehpur, and conduct my study on the bustards for the whole day, sometimes staying overnight in the village for two to three days at a stretch. When the Forest Department heard that we (myself and Bharat) were sheltering under a tree during the hot hours of the day, the CWLW of MP decided to build a hut for us near Fatehpur village.

After two months I left for Bombay to resume studies on the GIB in Nannaj, where the bird would breed in the monsoon in the grasslands. It was a hectic period of fieldwork for the next four to five months. When I returned to Karera in October 1982, a hut was ready for us. The construction left much to be desired, but it was far better than sitting under a tree! Someone named it the 'gypsy hut', and the name stuck! There was no water, electricity or toilet, but it was our home for the next five years. I even managed to get BHNS to sanction a princely sum of Rs 5,000 from the project fund for the construction of a toilet.

Here, I have to digress a little. The bustard is a large bird; males are nearly 2 m tall, while females are smaller, but still conspicuous. Many times, when I asked villagers working in their fields, '*Aap ne Sonchidiya deekhee hay yahan* (Have you seen the *Sonchidiya* in the area)?' most would reply with a 'no', even if the big birds were foraging in the distance and were within their field of view. In my arrogance, I would think that these villagers were dumb. How could they miss a 2 m tall bird? My presumption took a beating when I asked my local assistant, Munna, where could we get stone slabs for the toilet's roof. He said, '*Daktar Saab, Karera mey aap ke ghar key samne miltey hain* (They are sold in front of your rented house in Karera).' I said, 'Are you sure?' Munna replied, '*Jaa ke dekh leejaye* (Go and see for yourself).' He mumbled something about my ignorance, which I pretended not to have heard. Back in Karera, I saw that right in front of the window of our house, there were indeed some stone slabs piled up for sale! I realized then that if one is not interested in something – such as the bustard, or the stones in this case – they may not register its presence, large as it may be. We generally see what we want to see. Similarly, people who are not interested in birds often do not hear a bird's song, while a keen birdwatcher may hear it even against a noisy background!

As I wrote earlier, during my first visit to the Karera Bustard Sanctuary in 1981, I saw 3 bustards and was told that there were less than 10–12 birds left. In October 1982, I counted 14 bustards in the Sanctuary, along with two to three juveniles, which meant that there were at least 14 birds, with the possibility of a few more in the landscape outside the Sanctuary. So, I guessed there could be around 20–22 birds in the Karera landscape. The CWLW was very happy with these results, and the jump from the expected 10 birds to the 20–22 delighted him, even though that is too small a number for a viable population! I also found that the bustards start breeding from mid-March onwards, with display by two to three territorial males. Eggs are laid from April to June – the peak summer months. So, March to June became my main study period in Karera; after that, I would go to Nannaj and stay there from July to November. In the winter months, I wrote reports and papers, and conducted surveys in other states. Given the GIB's large distribution range, historically from Punjab–Haryana in the north to Tamil Nadu–Karnataka in the south, and Odisha in the east to Rajasthan–Gujarat in the west, the breeding season of the bird varies from region to region.

As I could not afford a house in Bombay with my meagre salary, J.C. Daniel allowed me to stay in the office, the formidable Hornbill House, whenever I visited the city between field sites. In Bombay, I would meet BNHS members, staff and visitors, attend BNHS functions, and sometimes go to Sanjay Gandhi National Park with J.C. Daniel and BNHS members. In the office, Mr Daniel had the reputation of being a strict disciplinarian (totally unfounded though!), but in the field he was quite different – joking, laughing, cajoling youngsters to learn more about nature and sharing food! Even the silliest questions from a novice would fail to rouse his temper.

Living in Fatehpur during different seasons, experiencing 48°C in summer and 1°C in winter, gave me immense strength and willpower, something I lacked earlier as a city boy. However, the best part of that tenure was that I learnt so much from the villagers – the importance of different seasons in their lives, their culture, customs, rituals, the importance of livestock, bonds based on trust, and their earthy logic and humour. But there was also caste discrimination, a life of utter poverty, fights for land and water, the status of women in different communities and the many strange cultural norms of a caste-fractured society. The last was visible in all facets of life. At that time, there was only one bus service in Karera. It would come from Sunari village in the morning, about 8 km from Fatehpur, taking passengers along the way, alerting them to its imminent arrival by blowing the horn loudly from 1–2 km away. The waiting passengers would be huddled in their own caste groups, but when the bus arrived, the male *thakurs* would enter first, no matter who was first in the queue, then their women, and lastly the other backward castes.

But some realizations were sweeter. We city people have lost touch with the lunar phases, but full moon nights are very different in villages, particularly in summer. One night in Fatehpur, I heard loud noises outside, so I asked Munna, '*Aaj itna shorgul kyon hain* (Why is it so noisy tonight)?' He said, '*Aaj chandni raat hai* (Today is a full moon night).' It was then that I realized how the dim light of a full moon can make a difference in the life of villagers – children were playing around, women were talking animatedly and elders were sitting outside, smoking hookah or *beedi*s till late at night. In contrast, on the 'dark' nights, there would be complete silence in the village, particularly in winter.

In the dry stony area of Karera, monsoon brought great joy, because it facilitated cultivation and the barren soil sprouted grass, which fed the hungry cattle. It was here that I first appreciated the pleasant smell of newly wetted parched soil, what we call petrichor in English. While in the cities we run inside when it rains, children and young girls in the villages go out to get soaked. This was another free lesson for a city-bred boy – to give importance to nature's cycles.

There was no medical facility in Fatehpur and its nearby villages, so a few times I had to rush sick people to Karera. The satisfaction of taking a sick or injured person to a hospital 20 km away compensated the loss of one day's fieldwork – bustard observation can wait but not a sick person. What I saw in Fatehpur, coupled with my inherent leftist leanings, strengthened the view that we needed radical changes to uplift our society. During my surveys in the Thar Desert, as well as in Maharashtra, Karnataka, Andhra Pradesh and Gujarat, I came across poor villages with almost zero basic facilities. Nothing can be more ironical than the fact that India builds world-class hospitals with state-of-the-art modern facilities in cities, while many primary health centres in MP or Rajasthan do not even have a clean syringe to take blood samples or basic medicines to treat a fever. Although my experiences date back to the early 1980s, I am not sure how much India has changed since then. A 2021 visit to a primary health centre in Lucknow for a Covid injection did not give me much assurance on this matter.

My old jalopy, which was given to me by J.C. Daniel, was an 18-year-old Willys Jeep, without a four-wheel drive. However, I took up the challenge and used it for 10 years, surveying the GIB (and later floricans) across the country, from Gujarat to Assam, and from UP to Tamil Nadu. As expected, it would give trouble regularly – I do not remember how many days we wasted in mechanic shops to get the old vehicle repaired. From

Rollapadu (Andhra Pradesh) to Sailana (MP) to Jaisalmer (Rajasthan), I got to know all the roadside mechanics. Some even became my friends. A few would say, '*Saab, aap phir aa gaye is khatara gaadi se* (Sir, you have come again in this old vehicle)?'

In 10 years of travelling in the old vehicle, I must have changed or repaired all its parts, including dismantling and reassembling her in Jhansi once. In 1985, when I took it to Jaisalmer, it stalled on the sand dunes thrice, between Phalodi and Jaisalmer town. Once, when I went to meet the District Forest Officer (DFO) of Barmer, the vehicle got stuck inside the town itself, on flimsy sand dunes. While talking to the young DFO at his office once we'd retrieved our jeep, I could make out that he was not convinced that we could survey the bustard areas of the Thar Desert in that vehicle! 'When a BNHS vehicle gets stuck inside the town, how will it go to the remote sandy areas?' he asked. My most embarrassing experience was in 1983, when our vehicle became stuck on a small sand dune between Sam and Sudasari in the Desert National Park. We had to request people from the Ganga village to rescue our vehicle. An impressive-looking middle-aged man said, '*Saab, isse purani gadi nahi mili thi* (You could not get a vehicle older than this)?' Even now, whenever I pass through that dune (much tamed now, thanks to the metalled road), his sarcasm reverberates in my ear.

However, in hindsight, I think my best days were in the 1980s, when I had the full freedom to do fieldwork all over India, in search of the GIB, Lesser and Bengal floricans, wetlands and grasslands. I thank 'Old Girl' for giving me so much (academic) satisfaction, by taking me around despite her ageing body parts!

I first met Ashish Chandola at Hornbill House, where he screened a documentary on the Pheasant-tailed Jacana (*Hydrophasianus chirurgus*), a bird he had filmed in Sri Lanka. He had earlier worked as an assistant to the famous German wildlife film-maker, the late G. Dieter Plage. *Jacana*, filmed in 1980, was Ashish's first independent documentary for Anglia Television (later ITV Anglia), which produced the long running and highly successful TV series, 'Survival'. In 1982–83, he filmed the Monal Pheasant in Nepal's Sagarmatha (Mount Everest) National Park, for a feature titled *Bird of Nine Colours*. This was screened for Dr Sálim Ali. During the screening, the great ornithologist, Ashish recalls, asked him to rewind the documentary, so that he could watch the courtship and mating rituals of the Monal again! The 'Old Man' later shared with him that the Monal did not make for good eating, unlike the Blood Pheasant!

In 1985, we (myself, Usha Lachungpa and Ravi Sankaran) met Ashish Chandola and Joanna van Gruisen in Manas National Park (Assam), where they were filming *Many Moods of Manas*, a documentary sponsored by Project Tiger. I have not seen a better documentary on this famous tiger reserve, which also featured the spectacular aerial display of the Bengal Florican, caught on film for the first time. Over discussions, we decided to make a documentary on the GIB. I invited them to Karera, and they eventually arrived in 1987. Cramped together in the small 'gypsy hut' on long, hot summer days, I came to know them closely. I admired their dedication in filming the bustards – who were luckily in full mating mode during those hot summer days – despite facing materialistic deprivations. Since summer days are quite long, filming would take place over two sessions – the first between daybreak and 10 a.m., and the second from 4 p.m. until darkness fell. Joanna would often sit for six to seven hours in a small hide, which I had made for her. Ashish would sit in another hide, at a spot where some bustards used to come for foraging. Later, we worked together to film the GIB in the Desert National Park. Ashish had bought a new gypsy, which had trouble starting, while my 'Old Girl', then 25 years old, behaved perfectly well (for a change!), perhaps to tease the new vehicle. We spent about a week in the Sudasari area of the Desert National Park, returning satisfied with the footage of bustards. Later, the couple went to Nannaj and Rollapadu for additional footage, all courtesy BNHS. Although nearly three decades old, it is still the best documentary on this rare bird.

During the BNHS Centenary Conference in December 1983, I met Pushp Kumar, the then CWLW of Andhra Pradesh, who told me that some bustards had been reported in Kurnool and Anantapur districts, and invited me to his state. December 1983 found me in the Rollapadu grasslands of Kurnool district, which turned out to be perfect for bustards. The dry grasslands were agriculturally unproductive, and so the state government had planned to distribute it to poor people; some houses had even been constructed there. But when bustards were found there, Pushp Kumar was keen to protect them. About 10–15 bustards were reported from the area. I saw 11 birds, five males and six females. With a view to studying them, I decided to repeat the visit.

So, in July 1984, I drove with Carl D'Silva, a young artist, from Solapur to Rollapadu for the second BNHS field visit. En route, I visited the Rannibennur Blackbuck Sanctuary in Karnataka, where we saw three female bustards. When we reached the grasslands of Rollapadu, it was raining heavily, so we took shelter in a temple. Once the rains stopped, we were greeted by the sight of 22–25 GIBs, in scattered groups, enjoying the feast of grasshoppers and alates of termites. I had never seen so many bustards together. We decided to stay in Rollapadu, but there was no place to stay. The kind temple priest allowed us to use the veranda, with strict instructions not to consume non-vegetarian food, not that we had too much food on hand to eat! The sight of bustards and frequent rains did not allow me to sleep. Getting up early the next day, we drove around the grassland and counted 35–37 bustards. The whole day was spent in studying the area. The following day, we rushed to Hyderabad to tell Pushp Kumar about our 'discovery'. He was elated and announced the news to the media, and the story (to use today's parlance) went viral across India. Soon after this, Pushp Kumar visited the area with his staff and a few members of the Birdwatchers Society of Andhra Pradesh (BSAP), and counted 40 bustards.

As the government had earmarked the so-called 'wasteland' of Rollapadu for distribution among the poor, Pushp Kumar asked me to meet the collector of Kurnool district and convince him to give the land to the Forest Department, on the grounds that a highly rare bird of India was thriving there. The Forest Department had already petitioned the government to transfer the land to them, but it wanted to put pressure on the collector, as he was the person on the ground who could do this. Plus, media reports of bustard sightings had created the right atmosphere in favour of the *Betta-meka Pakshi (Betta* is 'white' and *meka* means 'goat', in Telugu), the local name of the GIB. The young collector was quite supportive, and I also did a bit of name-dropping (Sálim Ali, Indira Gandhi) to further convince him that it would be a noble act on his part to provide the bird the grasslands area, where they could breed and propagate, and allocate some other land to the poor people. In any event, the undulating dry grasslands of Rollapadu (and surrounding areas) did not have a water source (except rainwater), so cultivation would have been difficult. The result of all this was that the then CMs of Andhra soon announced protection for the GIB's habitat in his state.

The credit for saving the bustard area in the early 1980s solely goes to Pushp Kumar, one of the most remarkable CWLWs that I have met during my interactions with the IFS. He was also brilliant at zoo design and management, as manifested in the 'open' zoos he created (a radical idea at that time) in Hyderabad and Visakhapatnam. For me, his greatest contribution was promoting a *fauj* (army) of young conservationists in his state. The late Siraj Taher, one of the founders of the BSAP, now renamed Deccan Birders, wrote in *Pitta*, the newsletter of BSAP in 2006, 'His association, guidance and help were the main reasons for the phenomenal growth of the society and its recognition all over the country.' To support the embryonic BSAP, Pushp Kumar provided transport to young birdwatchers, and would even accompany them on Sunday mornings.

With Pushp Kumar's support, our Bustard project was extended to Rollapadu, and in 1986 Ranjit Manakadan was posted there for a year. Another literal feather in the cap was the sighting of the Lesser Florican in Rollapadu; until then, the bird had been known to breed only in northwest MP, eastern Rajasthan and Gujarat. Blackbuck, the state animal of Andhra Pradesh, was also found in Rollapadu in small numbers, with poaching being the reason for their low numbers. In winter, hundreds of migratory larks were seen, and nearly a 100 harriers. A pack of wolves, too, made their appearance and decided to make the grasslands their permanent home. Newspapers were full of articles on the 'magic of Rollapadu's grasslands'. Members of the BSAP made a beeline to Rollapadu, each coming up with interesting stories. The photographic boom had not yet started, but we still managed to get remarkable photographs of Rollapadu's biodiversity. For the first time in the undivided state of Andhra Pradesh, the focus slightly shifted from the tiger to the GIB!

In 1991, when the Grassland Ecology Project began, Ranjit was again stationed at Rollapadu for three years. During the stint, he found out that a large canal project was coming up just 3 km from Rollapadu. We tried to stop the canal from being built near the grasslands, but the political and social pressure was so much that our efforts failed to have any impact. The canal and a large dam changed the landscape of the grasslands. The golden period of Rollapadu was short-lived. Today, only two or three female bustards survive in Rollapadu.

The fate of the GIB at Rollapadu symbolizes the downward trend of the species across India, including in its stronghold – the Desert National Park. Even in Rajasthan, according to WII researchers, around 100 bustards survive in two major pockets: about 40 birds in the Desert National Park, and slightly more in the 'field firing range of the Indian army', where they get protection in the nearly 2,800 sq. km that is under the army's control. Looking at the deteriorating condition of the GIB, since the 1990s, I had been urging the Government of India (GoI) to start a conservation breeding programme, for which I had many meetings but nothing was done for almost 20 years. During that span, I saw the sad extinction of bustards in Sorsan in Rajasthan, Karera and Ghatigaon in MP, and a drastic decrease in the Naliya grassland in Kutch (officially Kachchh) in Gujarat, Rollapadu in Andhra Pradesh, Nannaj in Solapur (Maharashtra), and many areas in the Thar Desert.

Fortunately, in 2019, the GoI, the Rajasthan state government and WII started a conservation breeding programme with the expert supervision of the International Houbara Foundation, UAE. As I write this in July 2024, I am happy to report that there are 43 bustards in two conservation breeding centres – Sam and Ramdevra – looked after by a dedicated team from WII (under the supervision of the highly talented Dr Sutirtha Dutta) and the Forest Department. At the same time, in situ conservation is also being done, so in a couple of years, when the breeding stock of 50-60 bustards is fully developed in the two breeding facilities, initially around 5–10 captive-bred birds will be released in areas where wild birds are found. This will be done every year while keeping the breeding stock intact. Minimum human contact will be ensured for birds that are to be released, and if necessary, these will be trained to live in the wild by providing them natural food. The well-known strategy of 'soft-release' will be executed so the released birds adapt to wild conditions.

The plan is to reintroduce the GIB in some of its former habitats, such as Nannaj, Rollapadu and Naliya. The reintroduction of birds will not be successful unless we keep large areas of their natural habitat intact. This is possible in the large 3,162 sq. km Desert National Park, as well as in the 2,800 sq. km area under the Indian army. The BNHS and WII have identified some more areas in Jaisalmer that are suitable for bustards – these need to be brought under the protection umbrella of some sort, such as conservation or community reserves. Regarding Nannaj, Rollapadu and Naliya, only time will tell how suitable they will remain in the next few years.

I do not know whether I will live to see the successful conservation of the GIB unfold. But '*Umeed pey dunya qayam hai* (The world lives on hope).'

7

Changes Come Calling

I joined BNHS in 1980, at a time of great change within the institution. From a small organization of around 500 members, a staff of 22–25 people and limited funds, it had grown into a team of 60–70. There was good inflow of funds and lots of visitors from the key funder – the USFWS. A prominent figure of the USFWS was David Ferguson – a great friend of Dr Sálim Ali and J.C. Daniel, and a keen communicator and listener. Ferguson kept a sharp eye on everything that was happening in the four projects that the USFWS funded in the early 1980s.

How BNHS got this funding bonanza is an interesting story. In the 1960s, when India was grappling with food scarcity, the US government came to its rescue under the PL-480 programme, or the Public Law 480. It was administered by the departments of state and agriculture and the International Cooperation Administration, and permitted the US president to authorize the shipment of surplus commodities to 'friendly' nations, either on concessional or grant terms. It also allowed the 'friendly' nations to purchase

US agricultural commodities with local currency, thus saving their foreign exchange reserves, while also relieving the US of grain surpluses. India decided to purchase the grains, but in view of the country's foreign exchange crises, the grain was paid for in Indian rupees. However, the US government had no use for the Indian currency and the money was kept in India for almost two decades, accumulating interest, even when India stopped buying grains after the Green Revolution.

At the end of the 1970s, the US government decided to use this accumulated fund, giving it to Indian institutions for research and infrastructure development in the areas of agriculture, medicine, engineering and poverty alleviation. As the environmental movement was gaining steam in India, it was decided that some funds would be given to conservation organizations as well. At that time, BNHS was India's most well-known conservation organization, and Dr Sálim Ali had worked with Dr Dillon Ripley, the secretary of the reputed US-based Smithsonian Institution, from 1964 to 1984. The USFWS, therefore, asked BNHS to submit proposals on wildlife conservation. The BNHS submitted four, thinking that one or two would get approved; however, to everyone's surprise, the USFWS agreed to fund all four!

Previously, BNHS used to get projects valued between Rs 5,000 and Rs 30,000, which would be executed by a few staff members, and members who volunteered in their spare time. So, when it received a huge grant from the USFWS, there was turmoil. All the projects had come in the name of Dr Sálim Ali as principal investigator. At that time J.C. Daniel, Dr Robert Grubh, S.A. Hussain and Naresh Chaturvedi constituted the scientific staff of BNHS. The big question was – how was BNHS going to handle these projects with four staff members? Some Executive Committee (EC) members suggested that BNHS execute the projects with the help of interested BNHS members, but Dr Sálim Ali, J.C. Daniel and USFWS did not agree, leading to an acrimonious fight between the two groups, one led by Dr Sálim Ali and another by Humayun Abdulali, the famous ornithologist, former honorary secretary of BNHS and member of the EC. Some EC members suggested that if BNHS did not have enough trained people, it should refuse the projects, but Dr Sálim Ali put his foot down and proposed that BNHS hire young staff and train them. The USFWS supported his suggestion. As a result, advertisements appeared in various newspapers in 1980 – including the one that I saw – which led to my entry into BNHS.

Suddenly there was an influx of nearly 30 researchers into BNHS, almost doubling the number of staff. There was no space in the office, but fortunately, all the new staff were posted directly to various field stations at Point Calimere, Bharatpur, Mudumulai, Karera and Nannaj; most new recruits only saw the Hornbill House during the interview. When I visited Bombay en route my two field stations, Karera in MP and Nannaj in Maharashtra, I could feel the palpable resentment of some BNHS members against the researchers; a big grouse was that Dr Sálim Ali and J.C. Daniel did not have time for them anymore, given the influx of new projects and staff. However, many other members were happy with the new changes; BNHS also encouraged interested members to visit field stations, and many did that, later sharing their experiences through slide shows at Hornbill House. Thanks to these presentations and interaction of members with visiting scientists of BNHS resentment against the new staff gradually decreased.

In effect, from an inward-looking, self-effacing and reticent organization, BNHS became a robust, dynamic and forward-looking organization, led by stalwarts such as Dr Sálim Ali and J.C. Daniel, along with the support of many EC members.

―

The year 1983 was a watershed year for BNHS – it turned a century old. Planning for the centenary started two years in advance. In his President's Letter in the first issue of *Hornbill* (1981), Dr Sálim Ali wrote, 'To celebrate the centenary of the Society in a befitting manner, several measures have been planned by the Executive Committee, some of which are already underway. The most important and enduring of these will be, it is hoped, a one-volume Encyclopaedia of Indian Natural History.' It was edited by Mr R.E. Hawkins, a member of the EC and editor of the Oxford University Press (OUP). Tall, thin, slightly bent, sporting an old pith helmet (of the kind that used to be popular with the British in India), Hawkins had also edited Jim Corbett's books, and was a great friend of Sálim Ali.

Year-long activities were planned, including an international conference, which was held at the Indian Institute of Technology, Bombay, from 6 to 10 December 1983. J.C. Daniel involved me a few days before the conference started, much to the chagrin of one my seniors at BNHS, who had hoped to marginalize me by making me in charge of the audio-visuals. At the start of each conference session, it was announced that speakers should contact Asad Rahmani before their presentations. I was also asked to sit in the front row, reserved for dignitaries, so that it was easy for me to access the stage, in case the speaker needed any assistance. Thus, I secured the privilege of sitting alongside people like Dr Sálim Ali, Sir Dillon Ripley, R.E. Hawkins, David Ferguson, Environment Minister Digvijay Singh, Lavkumar Khachar, Himmatsinhji, Dharmakumarsinhji and other legends of the conservation world. On the third day, someone asked me, 'Rahmani, you have been very active, but where is *he*? (name withheld). I have not seen him.' 'He' was the same senior person who had wanted to sideline me during the centenary celebration! I hope 'he' learnt the lesson: never demean anyone, lest it comes back to bite you one day!

Many young scientists made presentations in that centenary year, some for the first time in their life. Dr Sálim Ali and J.C. Daniel were the people who clapped the loudest at these presentations – the true leaders of the flock and true mentors. This did not go down well with some BNHS 'seniors', who lived in their arrogant world of 'seniority'. But the event was a great success – the first international conference of its kind organized at such a large scale in India by an NGO on its 100th birthday! It was also my first-ever conference, and gave me the confidence for public speaking, participating in discussions with experts and mingling with new people.

The conference gave impetus and new direction to BNHS researchers. Being a purely natural history organization, most of us were not familiar with advanced statistics and its use in analysing natural history observations. Through the conference, for the first time we learnt how to use statistics in ecology, telemetry and radio-tracking. In fact, everyone left the conference with some new information, new friendships and new collaborations.

Another bonus of the centenary celebration was the

publication of a compact edition of the *Handbook of the Birds of India and Pakistan* by Sálim Ali and Dillon Ripley. I was among those for whom the 10-volume set was beyond their means, but as soon as the compact edition came out, I bought it. I still consult it quite often, even though I now have a digital version of the 10-volume 'bible' of Indian ornithology.

As part of the centenary celebrations, on 15 September 1983, the foundation day of BNHS, Mrs Indira Gandhi, the then PM of India, spent almost the entire day attending the BNHS events. In the morning, she first visited Hornbill House, to meet a select group of people, particularly her friend (and her father's friend) Dr Sálim Ali. Four senior scientists were asked to make a five-minute presentation on BNHS projects, particularly the ones on the GIB, Keoladeo, Bird Ringing and Bird Hazards to Aircrafts. It's easy to deliver a 40–50 minute presentation on these domain subjects, but to encapsulate that in 5 minutes was a challenge. Dr Robert Grubh, Dr V.S. Vijayan, S.A. Hussain and I went through rigorous rehearsals with a timer. I remember J.C. Daniel and Prof P.V. Bole sitting in on many rehearsals until we perfected our timing. On the D-day, everything went smoothly in front of Mrs Gandhi. That very day, a veteran Congressman, Mufti Mohammad, had revolted against her leadership, but she appeared relaxed, enjoying the company of BNHS scientists and members. In the evening, she also graced a BNHS event at the National Centre for the Performing Arts (NCPA) on Bombay's Marine Drive. A key announcement at the occasion was Mrs Gandhi's gift of a 33 acre parcel of land to BNHS for establishing a Field Research Centre; the land was carved out of the 520 acre Film City adjoining the Sanjay Gandhi National Park.

Based on my 50 years of experience in the field of wildlife, I have no hesitation in writing that no other PM was as concerned about wildlife and environmental issues as Mrs Gandhi. An example of her genuine commitment to environment was reflected in 1972, when she was the only political head of a nation to attend the first international conference on environment at Stockholm, Sweden. The other was the host, Olaf Palme, the Swedish PM. I particularly liked what she said during her speech, 'One cannot be truly humane and civilized unless one looks upon not only all fellow-men but all creation with the eyes of a friend…'[17] The Stockholm conference, called the United Nations Conference on Human Environment, was a watershed in the international environmental movement, and put environmental issues on the international agenda for the first time. Twenty years later, when environment protection became a major concern, 57 heads of state, 31 heads of government, along with civil servants, conservationists, environmentalists, scientists and NGOs from 192 countries attended the Rio Conference in 1992. Great leaders like Indira Gandhi showed the way that others now follow.

A sadder memory of Mrs Gandhi dates back to the same period around the centenary year. In 1983, when I showed my pictures of the GIB (taken at Nannaj) to Dr Sálim Ali and J.C. Daniel, they were extremely delighted. At that time, those were probably the best pictures of the GIB taken in the wild. Despite his advancing age, Dr Sálim Ali wanted to film the display of the GIB, so I made arrangements for his trip to Nannaj. In 1984, the road journey was too long – a full day from Bombay to Solapur – so we travelled by the overnight train. In the morning, the entire Forest Department of the Solapur district received him at the station and took him to the Circuit House. I could hear murmurings among the staff about his age and fragile frame. After that arduous train journey, we all thought that he would prefer to take some rest, but within 30–40 minutes, he was sprightly and active, surprising the people gathered in the Circuit House. After a quick breakfast, the forest officers took us to Nannaj. Sálim Ali scanned the area and inspected the hide where he would sit. Since it was the end of October, the adult male bustard had almost stopped displaying, though it still hung around in its territory. I had placed the hide in such a way that if the male came to the display spot – which it sometimes did in the late evening – Sálim Ali would be able to film it. Nothing much happened that evening, so we decided to return the following morning.

The next morning, Sálim Ali ensconced himself back in the large hide. Due to his knee problem, I had placed a small stool inside, so he could sit. Hides for natural history observations and photography are generally small, but keeping in mind Dr Sálim Ali's weak health, I had made a large hide, so that he could manoeuvre and relax his tired muscles. He sat in the hide with binoculars and a tiny 16 mm film camera, through which he typically filmed; his 36 films are a proud possession of BNHS. The bustard came and foraged around the hide – and Sálim Ali was able to see the bird easily – but it did not display. After about 3 hours we escorted out 'Old Man', as he was affectionately called by his colleagues. While walking to the car, the flowering spear grass, which has spiky tips, stuck to his thick socks. Later, I remember how he sat on a chair outside the forest guest house, while we tenderly removed the spikes from his socks, after which the forest officers brought tea and biscuits. The DFO was keen to show Dr Ali the quality of his tree plantation at Barshi, about 58 km away. He took him away despite my protestations; I could see that the great ornithologist was tired. By the time he returned two hours later, we'd got the news that Mrs Indira Gandhi had been shot, and had been rushed to the All India Institute of Medical Sciences (AIIMS) in Delhi.

When I broke the tragic news to Dr Sálim Ali, he was crestfallen and kept silent for a long time, while we debated what to do next. The DFO suggested we return to the Circuit House so that Dr Ali could rest. The evening filming was, of course, cancelled. We started for Solapur, but on the outskirts of the town we encountered a mob, which was clearly angry over what had been done to Mrs Gandhi. The DFO decided that we should return to Nannaj, where he offered us a meal, but Dr. Sálim Ali refused to eat. We left him sitting in the shade inside the Nannaj forest guest house compound in a pensive mood – I had never seen him so sad. By 1 p.m., we heard on the radio that Indira Gandhi was no more. Until then, the news had been that she was fatally injured and that doctors were treating her. By 3 p.m., the DFO had arranged one more vehicle, along with armed guards, and we again started for the Circuit House. There was an undeclared strike in the bustling town of Solapur, but small groups of people were seen with garlanded pictures of Indira Gandhi. I was relieved when we finally reached our destination.

At the Circuit House, with great difficulty we convinced Sálim Ali to eat or at least have tea. He was booked to return to Bombay by Siddheshwar Express, accompanied by Ranjit Manakadan. We tried to persuade him to postpone his train journey but he was adamant. Moreover, he was sullen, sad and silent, so our entreaties had little effect. When we found out that many trains had been cancelled, as riots had broken out, we again pleaded with him not to travel. Fortunately, or unfortunately, the Siddheshwar Express was running on time, so he insisted on returning. I tried to telephone J.C. Daniel, but telephone lines were continuously busy that day. The next day, when J.C. Daniel came to know that Dr Sálim Ali was travelling by train in that fast-deteriorating situation, he was extremely upset with me. He was even more annoyed that I did not accompany him back to Bombay.

On reaching Bombay, Sálim Ali was quickly taken to his Bandra residence, where he remained unhappy for many days, for he had a great respect for Mrs Indira Gandhi. Her father, the legendary Pandit Jawaharlal Nehru, had given the second edition of Sálim Ali's *The Book of Indian Birds* to her as a birthday present; incidentally, that book had nurtured her interest in wildlife. Her passing was a great loss to BNHS and the conservation movement in India.

8

The Florican Project

In 1984, BNHS got another project on the two floricans found in India – the Lesser Florican and the Bengal Florican. Dr Sálim Ali and J.C. Daniel, who were very impressed with my work on the GIB, asked me to work on these bustards as well (floricans belong to the Bustard family). Both floricans, like most bustard species, are inhabitants of grasslands. The Lesser Florican (*Sypheotides indicus*) breeds in the monsoonal tropical grasslands of northwest (Gujarat, eastern Rajasthan, western MP) and peninsular India (Karnataka, Andhra Pradesh, Maharashtra). Earlier it was widespread, appearing in monsoonal grasslands for breeding and then disappearing to other parts of the country for the rest of the year, but now, it is a critically endangered species. Its larger cousin, the Bengal Florican (*Houbaropsis bengalensis*) lives in tall, wet grasslands of the Indian terai, southern Nepal and the floodplains of the Brahmaputra river in Assam and Arunachal Pradesh.

Before commencing work on the floricans, I read all the available literature on these birds. The International Council of Bird Protection (ICBP) survey reports were quite useful, but even more useful were the studies conducted by Dharmakumarsinhji, the bustard/florican guru, in the 1940s and 1950s. I met him at his residence near the Oval Maidan in South Bombay, where he showed me all his papers and photographs of the Lesser Florican, and talked about how he had ringed the birds to study their movement. 'Asad, find out where they go after the breeding season,' he told me. It took another 25–30 years – and the use of satellite tracking studies by BNHS, WII, Gujarat Forest Department and The Corbett Foundation (TCF) – to find the answer that Bapu (as Dharmakumarsinhji was affectionately addressed) had not been able to find, despite ringing 498 floricans between 1945 and 1950. Animal tracking technology developed during the 1970s and was perfected in the 1990s. The technology, which now uses satellite tracking as well, helped reveal that after breeding in the northwest during the monsoon, the Lesser Florican moves mostly to peninsular India; though it could be found anywhere, even in north India post-monsoon. It usually moves in search of grasslands, sometimes appearing even in Nannaj (Solapur) or Dudhwa in UP. In the last 2–3 years, BNHS, TCF and the WII have come up with wonderful results, which neither Bapu nor I could have ever anticipated.

Back then, based on a review of published literature on the Lesser Florican, mostly in the *JBNHS*, I created a map of the bird's distribution.[17] It helped me look for the bird in areas where it could be surviving. A meeting with Lavkumar Khachar and Shivrajkumar Khachar, who had conducted a few studies of the species in Jasdan and Rajkot during its breeding season, also greatly helped plan my surveys. In 1984, Usha Lachungpa was recruited for this project, and we commenced work at Sailana in the Ratlam district of MP. As J.C. Daniel did not want to send her alone with me, Meena Haribal, who later taught in the US for three decades, came along with us. On 30 August of that year, Usha, Meena, Mehboob, Ali Hussain and I set off. Ali Hussain is a famous bird trapper who was employed by BNHS, and Mehboob Alam, his eldest son, was then learning the methods of trapping birds from his father (more about them later).

Finding Sailana, a former principality, was easy, but finding a place to stay was not. For a few days, we stayed in the small government guest house, but when government officials arrived, we had to relocate to a decrepit, leaking forest guest house, with one large room and one bathroom. While the two girls were quite accommodating, I was a little embarrassed sharing a room with them. But work is work. Moreover, there was no alternative. Ali Hussain, Mehboob and the driver stayed with the forest guards. On 2 September, we saw the first Lesser Florican, a male, performing the characteristic jumping flight to entice the female. It was my first introduction to this lovely species, which has fascinated many birdwatchers and ornithologists, leading them to capture some remarkable images in recent years. We spent the whole month collecting observations, until the male floricans stopped jumping and started leaving the area. During our surveys of western MP and eastern Rajasthan that year, we saw 47 floricans while locals reported 85–90 floricans.

P.M. Lad, a forest officer whom I knew since the Karera days, came to Sailana and we spent many days in the field, surveying the grasslands of Ratlam, Dahod and Dhar districts. P.M. Lad was a unique individual – he had an elephantine memory, a colossal knowledge of wildlife, unimaginable energy, a stern attitude to staff, an intolerance for fools, but great affection for genuine wildlifers. He was also an excellent photographer. One memory from our bustard survey days stands out – it was either 1985 or 1986, and Lad was very upset with a corrupt forest officer (ranger) posted at the Ghatigaon Bustard Sanctuary (MP), who was giving him false reports regarding the number of bustards in the Sanctuary. Like a true forest officer, he was a doer. During a visit to Ghatigaon, when the ranger told him that bustards were now being seen in the Tigra area near Gwalior, Lad decided to walk through the forest – a distance of about 30 km – along with the ranger, frequently asking him questions about the wildlife found in the Sanctuary. The poor pot-bellied ranger would not have forgotten that punishment all his life! Whether he stopped being corrupt is a different matter.

In 1985, at the age of 89, Dr Sálim Ali showed great interest in visiting Sailana to ring floricans. He took a flight from Bombay to Indore, accompanied by his assistant Archana. The entire forest department, including P.M. Lad, went to the airport and received him with garlands – only the *band baja* was missing, but that was compensated for by the fawning staff and journalists. After breakfast in the forest rest house in Indore, Lad took Dr Ali in his vehicle to Sailana, while Archana and I followed in a separate vehicle.

We had been told that there was a man in Ujjain who could make good bird decoys, so Lad had asked me to find out about the quality of his work, as we wanted to use a decoy to attract wild floricans for catching and ringing them. So, Archana and I took a detour through Ujjain. In the crowded lanes of

Ujjain, it took us three hours to find the man, who immediately agreed to make florican decoys for a tidy sum. By the time we left Ujjain, it was already 3 p.m. and covering the distance of 135 km took five hours. And so, by the time we reached Sailana, Dr Sálim Ali had returned from the field. Not finding us at the forest guest house, he became worried. Meanwhile, we had first gone to the old rest house to freshen up, and by the time we met him, he was furious. That was the first time I witnessed Dr Sálim Ali's famous short temper, about which J.C. Daniel had warned me. However, he soon calmed down and we had dinner together. The decoy was not necessary the next day, as ace trapper Ali Hussain caught two male floricans on their territories. Both were ringed, measured and released.

In March 1985, at J.C. Daniel's suggestion, young Ravi Sankaran joined the Florican Project. Ravi was about 21 years old and had just finished his BSc degree from Loyola College in Chennai. The famous Catholic college, founded in 1925 and affiliated to the University of Madras, was known for its discipline and high-class teaching, though I found no evidence of this in Ravi!

Ravi was the wild child of a highly educated family – his father was a corporate honcho, and mother a social worker. Ravi spent his childhood in a boarding school in Switzerland, which failed to tame him. His maverick behaviour was a quality that I liked most about him. In fact, I've always liked people who are different from the hoi polloi. In Hindi we call such persons *bindaas*; however, Ravi's *bindaas* nature came along with sincerity in work, respect for elders, an ability to learn and improve, and the inner strength to be honest to the core.

For me, it was elder brotherly love at first meeting, despite Ravi's long, torn, dirty *kurta* over faded jeans, cheap flip-flops, three-day-old stubble and wild, curly unkept hair – in other words he was the perfect ragamuffin. He was the exact opposite of me – I was always meticulously dressed for office. Plus, his dreamy big eyes, under the shade of thick eyebrows, gave him the appearance of a drug-addict. I disliked his habit of shaking his legs and fiddling with his pen, and yet, something about him appealed to me. Being an iconoclast myself, I found him different from the usual staff. He was the epitome of what a young intellectual or a research scientist should be. Another quality in his favour – he was far removed from the societal prejudices that many upper-class Tamil Brahmins seemed to have.

J.C. Daniel placed Ravi under my tutelage, remarking mischievously, 'Try to control him.' A few days later, we were on the train to Jhansi, the nearest railway station to Karera. That 36-hour journey provided us ample time to get to know each other. His English was excellent and Hindi equally good, as he had lived in Bombay. After some time, I realized that he would disappear every 30–40 minutes. At first, I thought he was going to the bathroom, but when this recurred regularly, I sneaked up behind him, and caught the young man smoking outside the AC section! A mild reprimand was followed by shared laughter. I dislike smoking but allowed Ravi to smoke in front of me – a concession he always appreciated.

Two weeks of fieldwork in the heat of April in Karera tested Ravi's endurance. We sat together for hours on a small stony hillock to note bustard behaviour. It was their breeding time, so a lot of activity was going on around us. I taught Ravi to make field notes; back then, he was totally raw and could identify just four to five common birds. After two weeks, Usha and Ganden Lachungpa arrived in Karera to start work on the Bengal Florican. In mid-April, we four started for Dudhwa via Lucknow in our old jalopy. At that time, the road was so narrow and derelict that our vehicle took a full day to cover the distance of 238 km! It didn't help that our speed was 'controlled' by J.C. Daniel, who had asked the mechanic to install a speed controller before giving the old vehicle to me.

On 17 April, we saw our first Bengal Floricans – three adult males fighting in a grassland in the Kowaghati area of the Sathiana range. In my field diary I wrote, '3 Bengal floricans seen from 8.15 to 8.55 a.m., walking together on the grassland of Sathiana area near Kawwa-Ghati-pul. Swamp deer: males walked very close to the swamp deer, without any interaction.' Pardon the English of a field notebook. We spent a week in Dudhwa and then went eastwards to survey Katerniaghat, Suhelwa, Sohagi Barwa, and finally entered Bihar to visit the Valmik Tiger Reserve, and a few extant grasslands on the Bihar–Nepal border. From there, our destination was North Bengal and Assam. As the survey progressed, the notorious North Indian summer started heating up even more. While driving in Bihar, I got extremely tired. I then asked Ravi, 'Do you know driving?' He nodded his head in the affirmative. My next question was, 'Do you have a driving licence?' The nod was much stronger, followed by the remark, 'Should I show it to you?' Believing him, I stopped 'Old Girl' and gave over the wheel to him. For the rest of the three weeks, we shared driving duties. When we returned to Karera after five weeks, I asked Ravi to show me his driving licence. His matter-of-fact reply – 'I do not have it' – roused my anger against him for the first time. 'If there had been an accident, as a senior I would have been responsible,' I scolded. He replied with a blank expression, '*Kuch hua to nahi* (Nothing happened, you see)'. That was Ravi Sankaran for you.

We stayed friends for the next 25 years, until his untimely death in 2009 at the age of 46. In 1987, he decided to study Lesser and Bengal Floricans for his PhD, so we spent many weeks together studying the Lesser Florican during the monsoon in Sailana, and spent the summer in Dudhwa observing the Bengal Florican. There were plenty of memorable moments, but one rather hilarious incident is worth mentioning. During our first survey in 1985, we were staying in the Satsamali Rest House in Orang National Park (Assam) when a White-breasted Kingfisher (*Halcyon smyrnensis*) started calling. Ravi, a novice at that time, asked, 'What is this?' Just before that we had seen a large monitor lizard outside the rest house and so I jokingly said, '*Yeh monitor lizard ka bacha hai, maa ko bula raha hai* (It's the young one of a monitor lizard, calling its mother).' At that time, Usha, a good birder, came out of her room, so I moved away a little from where I had been standing. Ravi blurted out his new-found 'knowledge' to her. But one stern look from her revealed the joke! Later, in 2019, when I went to Satsamali and a White-breasted Kingfisher called, this old episode came to mind vividly, bringing tears to my eyes. Everything was the same – the old guest house, the railing on which one can lean and watch the rhinoceros and hog deer, the old mango tree, the kingfisher, the monitor lizard, but something – rather someone – was missing.

To study the behaviour and status of the Bengal Florican, Ravi worked for three to four years, from 1987 to 1990 in

Dudhwa, while Goutam Narayan (earlier from the Bird Hazard Project) and Lima Rosalind, Goutam's then wife and BNHS employee, worked in Manas (Assam). At that time (the late 1980s), Manas had the largest population of Bengal Florican in India. We all did extensive surveys in the UP terai and in Assam. Both Ravi and Goutam earned their doctorates on this amazing bird, which has now sadly declined to such an extent that the International Union for Conservation of Nature (IUCN) has included it in the 'Critically Endangered' category. While Lima, Goutam and Ravi shifted to other subjects and areas, I continued my survey and conservation work on the floricans. From 2013 to 2017, I ran two major projects on the Bengal Florican – one funded by the MoEFCC and another by the Preventing Extinction Programme (PEP) of BirdLife International.[18] Later, PEP funded another project on the threatened birds of the Brahmaputra floodplains.[19] During these surveys, I found that the Bengal Florican is still surviving in some protected areas of Assam and Arunachal Pradesh, with good populations in Kaziranga, Manas, Orang and D'Ering sanctuaries. Bombay Natural History Society and other organizations are continuing their studies and conservation work on this critically endangered species – which is much needed, lest our fragile grasslands lose one of their most charismatic species!

9

Desert Tales

For the love of bustards, I travelled to other countries as well. In 1983, the International Council for Bird Protection and the IUCN Bustard Group organized a survey expedition to Morocco. Paul Goriup, the chairman of the IUCN Bustard Group, invited me to join it. I quickly applied for a passport, mentioning my permanent address as Hornbill House (Colaba). Officers at the passport office were impressed that I lived in such a tony area of Bombay, not knowing that Hornbill House was an office!

That visit to Morocco was my first international trip; it was also the first time that I had ever travelled in an aeroplane. At that time, Indians were officially allowed to carry only US$15 on overseas trips. Living in my high moral world, I did not want to buy dollars illegally, but Dr A.N.D. Nanavati, honourable secretary of BNHS, and one of the finest human beings I have known, loaned me another US$20 (the amount that was left after his US trip). With the grand sum of US$35, I braved my first international trip!

Twenty days spent in the exotic country of Morocco, along with a team of British scientists, was a memorable experience.

Morocco had been under French rule, so locals could speak Arabic and French. In our 10-member team, one person could speak Arabic, French and English. Educated Moroccans speak French, so he would translate from French to English and back, but the local Bedouins with whom we spent lot of time in the Sahara, could only speak Arabic, so his job then was to translate Bedouin-Arabic dialect to English!

Despite living in a harsh desert environment, Bedouins are some of the most hospitable people in the world. During our search for the Houbara Bustard (*Chlamydotis undulata*) in the vast Moroccan Sahara, as soon as we came across a Bedouin camp or a shepherd's tent, we would be invited to have water and food. Refusing the offer is considered an insult, so our Moroccan host would invent all types of delightful excuses to refuse the offer. After getting information from the Bedouin about the Houbara's movements, we would leave quickly. During the month-long expedition, we could see only five to six Houbaras but found their tracks in many areas, indicating that the species was widespread despite the intense hunting. Based on the results of our expedition, we suggested the establishment of large community sanctuaries where hunting and shooting of Houbara and other wildlife would be prohibited. Subsequently, two major Houbara breeding centres were established from funds provided by rich Arab sheikhs; thousands of Houbaras are now bred every year at the Emirates Centre for Wildlife Propagation (ECWP) in Missour, Morocco, which was started in 1995 to primarily breed the North African Houbara, while a second breeding facility was established at Enjil, close to Missour, in 2006.

In 1986, I was invited, again by Paul Goriup, to attend a national meeting of the newly established National Commission for Wildlife Conservation and Development (NCWCD) in Saudi Arabia. Due to a visa issue, I could not attend the meeting but Prof. Abdulaziz Abu-Zinada, chairman of NCWCD, subsequently invited me in 1987 for a month's assignment as a consultant to the country's bustard conservation programme. This time the visa was easy, as I was a guest of the government. On landing at Riyadh, I was awestruck by the large glittering government buildings, wide roads, huge malls and hotels, but to my surprise the main headquarters of the NCWCD was in a nondescript building, which strongly reminded me of Indian forest offices. I do not understand why wildlife is a low priority all over the world, particularly in developing countries. The only saving grace was that some conservation work had begun in Saudi Arabia, which then was known only for its vast oil resources, camels, Bedouins and sand! After a preliminary meeting with Prof. Abdulaziz Abu-Zinada, a PhD in Botany and a respected scientist in Saudi Arabia, I was sent to Taif, where a National Wildlife Research Centre (NWRC) had been established in April 1986. Taif was also close to the Tihama region, the coastal belt along the Red Sea, where the Arabian Bustard (*Ardeotis arabs*) used to be found.

The Arabian Bustard was declared extinct in 1977 – a victim of extensive poaching. My challenge was to conduct surveys in the Saudi Tihama (some parts of the Tihama extend into Yemen, where we could not go) and try to find the bird. A young man named Mohammad Shobrak was assigned to me, and it was expected that I would train him in field studies. After the establishment of the NCWCD, Prof. Abu-Zinada recruited young BSc and MSc students (all male candidates) and assigned them to experts from European countries to be trained in wildlife fieldcraft; I was the only 'expert' from a developing country. Fortunately, I got Shobrak, an indefatigable youngster with a great interest in wildlife. We flew to Taif and drove to NWRC, about 30 km away, located at a height of 1,400 m.

The newly established NWRC, spread over 650 ha, had the mandate to breed the Houbara Bustard and Arabian Oryx (*Oryx leucoryx*). The fenced reserve was like a mini-French colony, as most of the researchers were from France. His Royal Highness Prince Saud-al Faizal, who initiated the Houbara breeding programme, had given the task to his French friend, who brought his own team of very talented scientists and vets. I was received with great respect initially, but when they came to know that Shobrak and I were going to work on the Arabian Bustard and make NWRC our base, there were some grumblings, which Prof. Abu-Zinada sorted out quickly.

On 19 October 1987, we started our survey in the Tihama in a brand-new BMW. Since I did not have an international driving licence, Shobrak did the driving during all my visits to Saudi Arabia. On the eighth day of our survey, we met a local Bedouin, Ali Mohammad, who confirmed having seen an Arabian Bustard a few days previously. He was able to take us to the spot where he had noticed bustard tracks. I was not satisfied with the tracks, as female Arabian Bustards tracks are similar to the large Houbara tracks. We needed to see the bird. After two more days of intensive fieldwork, on 29 October 1987, we sighted an Arabian Bustard in an area called Jabel Labiba. Our discovery proved that the bird was not extinct in Saudi Arabia, although it was clearly quite rare. The news that the Arabian Bustard was surviving in Saudi Arabia and that a 'doctor from India' had found it was played up in all the local newspapers! Prof. Abu-Zinada and Prince Saud were thrilled and talked to us through the walkie-talkie while we were in the field. I think I was the first Indian to travel to Saudi Arabia in search of a rare bird; we all know why most Indians go to the Middle East – to seek jobs and earn money!

After that initial success, I returned to India, leaving instructions with Shobrak to regularly survey the Tihama and find more bustards. The Tihama's long coastal belt extends from Bab-el Mandeb (Yemen) to the Gulf of Aqaba (Saudi Arabia). It is a rather narrow plain, spanning 80 km at its widest point, and home to many species of African origin, including the Arabian Bustard. The largest population of the Arabian Bustard is actually found in the Sahel region of Africa, but the bird is so named because the first specimen was perhaps collected for science from Arabia. The story goes that George Edwards, an English naturalist, made an illustration and wrote a description of a live bird kept at the London home of eighteenth-century naturalist and collector Hans Sloan. In 1758, the Swedish naturalist, Carlos Linneus, who invented the binominal nomenclature system, described the species in his book *Systema Naturae*.

The Tihama is wedged between the sea and the high escarp-ment mountains of Hejaz and Asir. The low foothills of the escarpment, which fringe the plain, can be considered as a distinct physiographic sub-region, the hilly Tihama. Many species of African origin, including the Arabian Bustard, are found in the Tihama. Hamerkop (*Scopus umbretta*), Bateleur

(*Terathopius ecaudatus*), Dark Chanting Goshawk (*Melierax metabates*), Helmeted Guineafowl (*Numida meleagris*) are some species of African descent that were 'lifers' for me. Over the next couple of years, I returned to Saudi Arabia four more times, to guide Shobrak and to develop a conservation strategy for the Arabian Bustard and other wildlife of this fascinating region – also the greenest – in that vast, oil rich country.

A Tihama bird that interested me greatly was the Greater Hoopoe Lark (*Alaemon alaudipes*). I do not believe in God but would say that Nature (or God, if you are a believer) has banished this poor bird to the most desolate and barren areas even by desert standards. In the Tihama, too, it was found only in the super-arid areas, not the fertile plains. I remember how, in March 1992, after 15 days of futile search for the Arabian Bustard, when the ambient temperature was nearly 46°C, and the surface temperature closer to 50°C, I dozed off in the air-conditioned SUV, while Shobrak was driving, listening to his favourite Bonny-M songs. Suddenly, a bird darted in front of our vehicle, running briskly like a ballerina on the hot sand, looking for insects. I asked Shobrak to stop the car; we watched in admiration the adaptability of the Greater Hoopoe Lark, a quality that has allowed it to survive in the super-arid region. This was not my first meeting with this remarkable bird – I had seen it in the Moroccan Sahara. Since that first sighting, I had been fascinated by the bird, and every encounter brought me joy. (Later, I saw many Greater Hoopoe Larks in the Thar Desert and the Great Rann of Kutch.[20]) I advised Shobrak to carry out studies on the bird for his PhD, as we did not know much about this species that is distributed from the Moroccan Sahara, through north Africa, to the Middle East and the Thar Desert.

Shobrak went on to work with Joseph B. Williams and B.I. Tieliman in the NWRC, and brought out many papers on this species. What they found was fascinating – sometimes, the desert heat gets too high even for the Greater Hoopoe Lark, with surface temperature touching 62°C in its habitat! On such occasions, it enters into the burrows of the herbivorous Spiny-tailed Lizard (*Uromastyx aegyptia*) to escape the heat. Since the lizard is vegetarian, the bird has no fear of being preyed upon, and like a good neighbour, the Spiny-tail shares its home with a friend in distress. In a research paper published in *The Condor* journal in 1999, Shobrak and other scientists proved that the Hoopoe Larks can potentially reduce water loss by as much as 80 per cent by sheltering in Spiny-tail burrows during the hottest periods of the summer days. They also found that Dunn's Larks (*Eremalauda dunni*), Bar-tailed Desert Larks (*Ammomanes cincturus*) and Black-crowned Finch-larks (*Eremopterus nigriceps*) also use burrows as thermal refugia during hot summer days in the Arabian desert.

As the Arabian Bustard was extremely rare, it was difficult to sustain our interest in it, particularly when we were working in a harsh environment from sunrise to sunset, with only a dim hope of seeing the bustard. To keep ourselves busy, we took notes on all the birds we saw in the Tihama. Collating our field data from the six or seven trips between 1987 and 1992, we published a paper titled 'Birds of the Tihama Coastal Plains of Saudi Arabia' in the *Ornithological Society of the Middle East Bulletin* in 1994.[21]

Prof. Abu-Zinada invited me to work for NCWCD but

I politely refused, as I was committed to BNHS for life. However, I did agree to travel to Saudi Arabia on short visits until Shobrak was fully trained. I encouraged him to enrol for a PhD in a good university and work on the Arabian Bustard or Hoopoe Lark. As sightings of the Arabian Bustard were erratic, he finally completed his doctoral studies on the ecology of the Lappet-faced Vulture (*Torgos tracheliotus*) from Glasgow University in 1996. After the French left the NWRC, he took charge of the Centre, where he worked for 10 years, before going on to teach at Taif University. Mohammad Shobrak also represented the Middle East region on the Global Council of BirdLife International. I am as proud of him, as I am of my other PhD students.

10

AMU Revisited

I left BNHS in 1991 to join AMU's Centre of Wildlife. It was a difficult decision, one that I never wanted to make. I loved my work on bustards, floricans, grassland and wetland birds, as well as the working environment of BNHS. However, around 1989–1990, differences had started cropping up between me and J.C. Daniel regarding the working of my juniors. I was deeply hurt, because instead of trusting me and reprimanding my assistant, he took the latter's side. I felt humiliated and had to put my foot down aggressively, a course of action he may have perceived as misbehaviour and arrogance, though ironically, I respected him very much.

This incident had been building up for a while. Though J.C. Daniel was a very kind-hearted person and would always support the juniors, he would never check who was really at fault. This had happened with all three of my senior colleagues – Dr V.S. Vijayan, S.A. Hussain and Dr Robert Grubh – leading to a big fight each time. What J.C. Daniel did not understand was that juniors could be at fault, too, by not performing to the standards of their superiors. Later, I realized that he was even retaining under-performing staff, as he just could not bring himself to throw out anyone. He would first agree to take action against such staff, but a few tears in his chamber would change the situation completely. In retrospect, I think that he was too kind-hearted a gentleman to see anyone in distress. But at that point, I couldn't summon this perspective and hence my outbursts against him. I still feel guilty about my behaviour. When I rejoined BNHS as director in 1997, we would travel together in the BNHS car to the office (his residence was en route), and if I remember correctly, I apologized to him thrice. He would affectionately hold my hand and say, 'Rahmani, forget it.' It was his greatness that he forgave me, but I have not forgiven myself for my error of judgement – it still hurts like a spear in my heart, today and maybe forever.

When I joined AMU's Centre of Wildlife (later to become a Department) as a reader in 1991, it was housed in three rooms, in a corner of the impressive Sir Syed Hall. Trading the large rooms and tall windows of Hornbill House for a small stuffy room seemed like a great fall for me. The three-room office in Sir Syed Hall building also did not fit Prof. Musavi's vision for the Centre of Wildlife. So, he wanted to relocate the Centre to Baitul-Islam *kothi*, which had eight rooms, a large lawn and a courtyard. Prof. Musavi had been allocated an official residence, but he kept the *kothi* in his possession, paying a penalty for a year from his pocket, until he was able to convince the university authorities to allocate the *kothi* for the Centre. How many people can do such things to fulfil their dreams?

Baitul-Islam *kothi* was a typical old residential building, with interconnected rooms, three bathrooms, verandas, a high-walled courtyard and lawns on two sides. People laughed when he got permission from the University Building Department to convert it into a wildlife centre. But Prof. Musavi was quite innovative, and was able to convert the bedrooms into classrooms and the lobby into a library! Other inside rooms were converted into laboratories and computer rooms; I was allotted a small 8 by 8 m room. The large courtyard with guava, date and pomegranate trees provided some greenery, but no fruits.

The only redeeming feature of those early days was the presence of students. Their energy and enthusiasm allowed me to gradually forget the Bombay days. After some serious thought on the situation, I came to the conclusion that the only way forward was to enjoy teaching, reading and writing. Gradually, I began to love teaching, particularly wildlife ecology and biogeography. Every year, we would get a new batch of students, mainly from UP, Bihar, Kashmir and a few from the other states. All of them were young and enthusiastic, but most did not know much about wildlife (except Dhananjai Katju and Rajat Bhargava), so I had to start from the basics. Thanks to our pedagogical teaching in schools and colleges, most students had limited general knowledge. Without knowing the geography of the country/world, it is not easy to explain animal and plant distribution, their ecology, endemism and environment. So, I designed a full course on biogeography. Large maps of India and the world were displayed in the classroom, much like in kindergarten. I explained to the students that without having a good knowledge of rivers, mountains, geology, climate, weather, rainfall and temperature patterns, forest types and local communities, it would be impossible to understand the distribution of wild animals.

Teaching was hard work but also a great way to gain knowledge. I studied many books to develop the course that I first taught at AMU. At that time, there were not many books on Indian wildlife and ecology. Prof. M.S. Mani's book *Ecology and Biogeography in India* (1974) and Rodgers and Panwar's *Planning a Protected Area Network in India* (1988) were of great help, along with others from Prof. Musavi's collection. I would spend four to five hours preparing for an hour-long class.

I also tried to inculcate reading habits among students. The Centre was well stocked with newspapers, and sometimes, before the class I would quiz students on the headlines of the day. My maverick behaviour would surprise some, and amuse the others! Teachers were supposed to teach the course, not ask political or literary questions. Prof. Musavi had developed a good library at the Centre, and had also donated his personal books. I kept an eye on who was reading what and for how long, by regularly checking the library register. Then I would surprise students by asking them questions on the book they had borrowed. When I meet my former students, even 20–25 years later, they recall my insistence on reading.

Prof. Musavi understood my love for fieldwork. So, he provided a vehicle to take students to Sheikha Jheel, Ashpan wetland, Gursikaran grassland/forest and other nearby areas to learn about birds, blackbuck, nilgai, wetlands, grasslands and plantations. The students, too, enjoyed going to the field. Field visits became more frequent when I became Chairman of the Department. Many times, I deliberately took them to the field on a Friday afternoon, just after the *namaz*. There is an old mosque near the Centre, and just as I would be driving out to the field, the *namazis* would be leaving the mosque after their prayers. As there would often be female students seated beside me in the front passenger seat, the sight would shock the fundamentalists among them! But my choice of day and time was deliberate – the idea was to send out a message that both male and female students were equal. Soon, news spread in the University that the Chairman of the Wildlife Centre took girls (not students, mind you) out for *ghoomane* (roaming around). Fortunately, the dean in the Faculty of Life Sciences, first Prof. Man Mohan and later Prof. Shamim Jairajpuri, were supportive of our work. Incidentally, wives of both the professors were scientists and teachers in the University. I believe in women's empowerment and freedom – this is what I have always practised and also evangelize. I have nothing but disdain for people who do not give total freedom to women in the name of religion or tradition.

Murmurs were also heard when two young external researchers, Sarala Kaling and Sunita Pradhan, working on the Satyr Tragopan (*Tragopan satyra*) and Red Panda (*Ailurus fulgens*), respectively, in Darjeeling, came to AMU in 1996 for a year to write their thesis. By then, I was the Centre's chairman and ensured that all necessary facilities/infrastructure were provided to them, despite the initial buzz over encouraging 'outsiders'. One of my aims for doing so was to make a point to AMU's female Muslim students – that when two young girls could come all the way from West Bengal, they too could fight for their *azadi*, and the right to do what they wanted to with their lives. Capitulation to fundamentalists is not an option for young women.

At that time, we did not have many computers at the

Centre, therefore, time slots were allocated to researchers, starting from 7 a.m. until 10 p.m. Thanks to the frequent power cuts back then, the schedule would go topsy-turvy, leading to bickering among students. On top of this, two 'outsiders' had joined the pool. Sometimes these two girls and other female researchers would leave the Centre late, by 9 or 10 p.m., leading to rumours in the campus, but who cares when '*dil saaf hai* (when the heart is pure)'. So, when I received a circular from the proctor that girl students had to be back in their hostels by 7 p.m., I tossed it in the *raddi ki tokri* (waste paper bin), where such a diktat deserves to be. I wrote back to the University stating that since I too worked late until 10 p.m., there should be no problem for girl students to work late in the Department, and that if they had any doubts about my integrity, they should inform me. I never got a reply.

I am sad to see that from being one of the most progressive universities in India, AMU is now under the grip of fundamentalists and mullahs. In the 1970s, and even up to the 1990s, girls were seen without burka or hijab, but now around 30–40 per cent of girl students can be seen wearing a hijab and some even a burka. What a great failure of Sir Syed Ahmad's dreams to make his institution like Oxford and Cambridge! I now call AMU a 'glorified madrasa'. Is this the same university that had a vibrant culture of drama, music, discourse and debates on various subjects? Unfortunately, such deterioration is seen across all the universities lately, where fundamentalists and communalists are waiting to silence any discordant voice. I believe that universities should be the centres for discourses, dissenting views, new thinking, questioning of the traditional tenets and a little youthful radicalism. As a famous maxim goes, 'If everyone thinks alike, no one thinks.' We need universities to produce future leaders, thinkers, social reformers and scientists with new ideas. Respecting elders and teachers is fine, but this should not come in the way of new ideas, new philosophies and ways of re-examining existing norms.

It took me about a year to adjust to the Aligarian pace of life. Though I enjoyed teaching, it was not giving me time to go to the field to conduct surveys and studies, nor did I have the projects for fieldwork. I led a comfortable life in Aligarh, with a good salary, but my golden days in Karera, Nannaj and Rollapadu, and other field memories haunted me. Desperate to revive that dream, I began looking for a new job related to wildlife field research, or better still, a way to return to BNHS. J.C. Daniel had retired in 1991 and Dr Jay Samant had taken over; he would hold that post until 1996. I knew Dr Samant from the time he was involved in WWF-India activities, while teaching at Shivaji University in Kolhapur. When he asked me to come back to BNHS, I jumped at the opportunity. I was keen to supervise the Grassland Ecology Project, which I had left when I quit BNHS. My friend, Dr Vibhu Prakash who was at Bharatpur, booked a ticket for me to Bombay. After getting a four-day leave, I quietly left Aligarh for Bharatpur, but my bus had a puncture, which led to a delay. Meanwhile Vibhu was waiting for me at the Bharatpur station with my ticket, but by the time I reached, the train had left. For a change, it was running on time!

Missing the train seemed to relay a message – do not escape the current situation but fight it out! I decided to do it. The Grassland Ecology Project, which I had proposed and got funds for from the USFWS, was wobbling without a leader. Dr Samant could not give much time to it, as he was busy with his own administrative and research work; most of the senior scientists had left BNHS while Dr Vibhu Prakash was busy with the Raptor Ecology Project. So, Dr Samant suggested that I guide the project from Aligarh. A visit to BNHS along with Prof. Musavi in 1993 clinched the agreement. Although the staff was paid by BNHS, some funds and a Gypsy jeep were transferred to Aligarh.

Finally, I was back in the field! I was able to go to Solapur, where Satish Kumar (who was selected in BNHS when I was still there) was working on the Indian Grey Wolf (*Canis lupus*); to Rollapadu, where Ranjit was studying the grasslands, bustards and floricans; to Kutch, where Jugal Kishor Tiwari of BNHS was working on Banni grassland birds; and to Dudhwa, where Salim Javed (then my student) was working on birds. Though it was not happening at the pace that I was used to, fieldwork was a welcome respite from the drudgery of AMU.

Meanwhile, I got a small grant from the Oriental Bird Club to study the Stoliczka's Bushchat (*Saxicola macrorhynchus*), now known as the White-browed Bushchat. The genesis of this study is interesting. While developing a database of the birds of arid areas and grasslands, I went back to the *Handbook of the Birds of India and Pakistan* by Sálim Ali and Dillon Ripley – the ornithological bible of India. I was intrigued to read that this arid-zone bird, Stoliczka's Bushchat, first described to science from a specimen collected in Rapar, Kutch, was reported from Aligarh. I had seen some dry areas on the Aligarh–Mathura road, so I went there in search of this small bird, but could not find it. I also realized that during all my surveys in the Thar Desert, I had never seen this bird. So, what happened to this species? In 1991–92, my friend Thomas J. Roberts, who worked in Pakistan for 28 years, brought out the two-volume *Birds of Pakistan*, published by OUP. In the book, Roberts wrote that this bird had become extinct in Pakistan. Why and how? Its habitat in the Thar Desert of India and Pakistan has not changed much, and why would anyone kill this small sparrow-sized bird? It is not a game bird or even a cage bird. If an experienced person like Roberts considers it extinct in Pakistan, there must be some merit to it, I thought. I decided to find out the reason.

I collated all the old sight records from India and Pakistan, and found that the only recent record from India was by a Dutch ornithologist, from the Khara area on the Jaisalmer road. In 1993, I published a small note in the *Oriental Bird Club Bulletin*, titled the 'Little Known Bird: White-browed Bushchat'.[22] Based on this, the Oriental Bird Club asked me to conduct a proper survey, for which they gave Rs 35,000, a rather big sum at that time. Simultaneously, I had submitted a project proposal to WWF-India to study the fauna of the Thar Desert, which was approved. Having a total of nearly Rs 1,00,000 in hand, I conducted four extensive surveys in Thar Desert and Kutch over 1993–94. In every survey, I would take one student from the Centre, in addition to Mehboob Alam, my assistant. We found that the bird was not extinct but thinly distributed in the desert, and easily confused with other species.

These surveys resulted in a few research papers and a detailed report. World Wildlife Fund-India was so impressed by the report that they agreed to print it as a book. This led to my very first book, *Wildlife in the Thar*, which came out in 1997. Looking back, I feel that I could have analysed the data in a better way, but being my first book, it has a special place in my life. My paper, 'Status and distribution of White-browed Bushchat in India', published in the British journal *Forktail*[23] created interest in this bird; now, it is one of the species that is on the bucket list of birdwatchers visiting the Thar Desert.

Unfortunately, we still do not know much about its movement and breeding sites. Even its nest, clutch size, eggs, breeding success and threats are unknown. Similar to this small bird, there are hundreds of other species we do not know much about, while crores of rupees are spent on studying the 'ecology, food, and dispersal of the tiger'! India is so tiger-centric that, sometimes, it looks like no other species matter.

To return to the subject of the Grassland Ecology Project, what seemed like a beautiful solution (the execution of a BNHS project via AMU) unravelled in 1993–94; the dual authority created indiscipline and problems in the execution of the Project. In October 1995, the Project staff were called to Aligarh to analyse the data and write a report. Except for a few chapters, I was not happy with my own report; nevertheless, we had to submit it to the funding agency. Owing to various reasons, I was neither in full control of the Project nor of the BNHS staff working on it. The only silver lining was the excellent pioneering work done by Satish on the Indian Grey Wolf (*Canis lupus*) at Nannaj and surrounding areas, and by Salim Javed on bird communities in Dudhwa, including a study on the Red Junglefowl (*Gallus gallus murghi*). Based on these studies, we published many papers (too many to list here). Prof. Mark Behan, emeritus scientist of Montana University's (US) Botany Department, was our external advisor, and quite popular among the staff. On each of his trips, I arranged his lectures in the Centre and even at BNHS.

Before leaving BNHS in 1991 I had submitted another proposal for a Stork Ecology Project to the USFWS for funding. It was accepted and in 1994 the USFWS decided to fund it through AMU, as I had shifted there. It was a major project, given the Centre's 'standards', so Prof. Musavi was quite happy. I wanted studies to be conducted on all resident storks of India, primarily across three sites – Bharatpur, Dudhwa and the Assam valley. Gopinathan Maheswaran, Farah Ishtiaq and Hilloljyoti Sangha were recruited for the project, and all the three did excellent work. Farah worked on the comparative breeding ecology of the storks in Keoladeo National Park, Bharatpur, and highlighted that how the four stork species – Painted Stork (*Mycteria leucocephala*), Asian Openbill (*Anastomus oscitans*), White-necked or Woolly-necked Stork (*Ciconia episcopus*) and Black-necked Storks (*Ephippiorhynchus asiaticus*) – partitioned the same resources (fish, frogs, snakes) through space and time. Maheswaran's studies on the breeding ecology of the Black-necked Stork in Dudhwa National Park highlighted the importance of natural wetlands for this large, territorial bird that depends on large fish. We found that the Black-necked Stork is a good indicator of a wetlands' health. Hillol's work on the Greater Adjutant (*Leptoptilos dubius*) in Assam, Nagaon town to be precise, also proved the importance of natural wetland protection, along with fish resources. Though the Greater Adjutant can feed on offal around slaughter houses, its primary food for chicks (such as fish, snakes, frogs) is gathered from wetlands. Wherever wetlands and tall trees survive, as was the case in Nagaon district, it can breed on tall trees growing among human dwellings.

Having learnt a lesson in the Grassland Ecology Project, I promised myself that I would give more attention to the Stork Project. As I had selected the students, the whole responsibility of Project execution was on me. Fortunately, all three students performed well and the USFWS was happy with the report and research papers. In total, we published 25–26 research papers on storks in some of the finest subject journals of the world and of India.

Prof. Musavi retired in 1994 and I became the chairman of the Centre of Wildlife. I got busy with teaching, fieldwork, managing two major projects on grasslands and storks, not to mention the interaction with the University's bureaucracy, which hardly left me any time to read or write. In 1995, I fell sick – a large tumour was detected on the prostate. I decided to go to Mumbai to get it tested at the Tata Cancer Hospital. Dr Renee Borges, who was working in BNHS then, helped me a lot, as her elder sister was a cancer specialist. She and Dr Vibhu Prakash took me to the hospital, and like family members, remained there till the biopsy was done. Then, the worrisome wait started. What if I tested positive? Was the end so close? Renee broke the news that it was a benign tumour, but needed to be surgically removed. Renee's smiling face while she delivered the news will forever remain in my eyes. In my large BNHS family, she was like a younger sister, and younger sisters are known to bring joy. She did. To me, one of the best Indian festivals is Raksha Bandhan. I think it should become a global festival, like Christmas. I underwent surgery at the Jawaharlal Nehru Medical College, AMU. It kept me out of action for 15–20 days, with my colleagues and students taking care of me all the time. I was their 'family' patriarch!

One of the highlights of working in a university are the people we get to meet. One such encounter, which fostered a life-long friendship, was a meeting with Subodh Nandan Sharma in early 1992. Subodh sahib became interested in wildlife and nature when he was posted at Dhikala (Corbett Tiger Reserve) from 1965 to 1974 during the construction of the Ramganga Dam. After retirement, Subodh sahib dedicated his life to conservation. He established the Hareetima Environmental Group, and regularly publishes a newsletter in Hindi, named *Hamari Dharti*, from his own resources. Despite limited means and advancing years, he still visits different wetlands. During my stint in AMU, we frequently birded together – Sheikha Jheel, Aama-Madhapur, Rati-ka-Nagla and Patna Jheel became our common destinations, particularly in winter. In 1997, we wrote the *Management Plan for Sheikha Jheel*, which was jointly published by the Centre of Wildlife and the Hareetima Environmental Group. We continue to stay in touch. Subodh sahib is a real ground worker, a little-known brick in the edifice of India's conservation movement.

Some meetings are momentary, but just as momentous. In January 1997, when I heard that Medha Patkar, the famous activist and the main voice of Narmada Bachao Andolan, was in Aligarh, I invited her to deliver a lecture and interact with students. She readily agreed. Considering her popularity, I knew that she would attract a large crowd. Later, even without much publicity for the talk, the lecture hall was bursting with students, with half of them standing and many peeping in from the windows. Medha gave an eloquent lecture in *shudh* Hindi, followed by a standing ovation of 5–10 minutes. After the talk, she was mobbed by students and had to be 'rescued'! I was proud that I had been able to bring such a famous personality to the University. In her khadi sari, chappals, unkempt hair and a large bindi, she represented the quintessential Indian simplicity – a simplicity and culture all Indians should be proud of. She is not popular with the new Indian government, but I am sure she will be fondly remembered in history, while her critics will disappear in the anonymity of time.

Even though leaving BNHS and joining the Centre of Wildlife was not the happiest decision of my life, in hindsight, I think it was the right thing to do at that time. There were many reasons for this – some can be written while some others will remain unsaid, as the progress of time has blurred them. In the early 1990s, BNHS was in turmoil, as most of its major projects were over (except for my Grassland Ecology Project and Vibhu Prakash's Raptor Ecology Project), leaving 20–25 trained scientists in the lurch. The newly created Sálim Ali Centre of Ornithology and Natural History (SACON) absorbed some scientists – the best ones – but there were many more to be accommodated. As a result, some were shifted to my project, without my consultation, and others to the Raptor Project. And so, both Vibhu and I had to work with people who were not the right fit for our projects. I found an opportunity to leave, which I took, but Vibhu suffered for many years.

As I had written earlier, for the initial one year or so, I had difficulty in adjusting to the Aligarian way of life, but eventually, I made peace with my destiny. The raw energy of students and the pleasure of sharing knowledge with them buffered the drudgery of administrative work. By 1992–93, I could go to the field, not as often as I wanted to, but nonetheless on a regular basis. I also began to love teaching. But the most satisfying part was to see my students grow from callow, naïve and inexperienced fellows to dedicated defenders of wildlife and environment. I am still in touch with most of my students and researchers from the AMU days, including Farah Ishtiaq, Hilloljyoti Singha, Satish Kumar, Intesar Suhail, Khursheed, Rajat Bhargava, Dhananjai Mohan, Ifshan Dewan, Maqbool Baba, Imtiaz Lone and many more. They are the new leaders of the conservation movement in India. Did I contribute to making them so? I do not know, but that is my hope.

11

Back at BNHS

In July 1996, Vibhu telephoned to tell me that Dr Jay Samant had resigned (he was rejoining Shivaji University), and that I should apply for the post of director of BNHS. These were the sweetest words that I had heard from a friend in many years. The Grassland Ecology Project was over, and the Stork Project was coming to an end too. And so, I was not keen to continue in the University – it seemed like the perfect time to apply to BNHS. In November 1996, I was selected as director, BNHS, but could only take charge at the Hornbill House in May 1997, after completing my academic commitments at AMU.

As I had worked in BNHS for almost 12 years, and it had never left my DNA and mind even when I was in Aligarh, re-adjusting to the BNHS working ethos was easy. However, I now had a new role. But I eased in smoothly, thanks to full cooperation from J.C. Daniel, popularly known as 'J.C.' among his colleagues, BNHS President B.G. Deshmukh, Honorary Treasurer Sunil Zaveri, other EC members and the staff. My four-year chairmanship at the Centre of Wildlife (from 1994 to mid-1997) helped me learn the nitty-gritties of administration, albeit of a slightly different type. Also, I had continued to lead an active life in Aligarh, so a fast-paced job was not a problem after rejoining BNHS.

To understand what it takes to run BNHS, one must first understand how it is organized. It is a membership-driven organization and every four years, members elect 12 representatives to the EC, now called the Governing Council (GC). The 12 elected members nominate a president and three vice presidents, either from within the elected group, or request 'outsiders', typically public figures, to assume these influential posts. The EC meets every quarter, sometimes even five or six times a year, so the director has the help of an honorary secretary (HS) to set the agenda.

Being an NGO with no fixed funding, the BNHS director has to be an expert in fundraising and bringing in research projects. As can be expected of a scientific organization, publications are important – so the top boss has to see to it that the *Journal* (Journal of the Bombay Natural History Society) and *Hornbill*, which is a popular magazine, are edited properly and come out on time. Members' activities and conservation education put additional pressure on the director's time, along with representing BNHS at various government committees. On top of this, the director has to be a scientist – so he (there hasn't been a female director yet) has to write project proposals, deal with funding agencies, get funding, supervise projects, write reports and papers, or in case projects are being executed by other senior scientists, ensure that reports/papers are submitted on time. And of course there is the fiscal management to take care of! Also, thanks to the explosion of digital communication, the director receives hundreds of electronic messages (emails, texts) each day, and all need to be responded to.

Luckily, I didn't have to do all of these by myself. Institutions are built and nourished by people, and as American writer Shire Hite aptly said, 'People make institutions, not vice versa.' This is true of BNHS too. In its long history, BNHS has had many stalwarts who dedicated their lives to the Society and natural history studies. I was lucky enough to work with some of them.

When I joined BNHS as its director, B.G. Deshmukh was the president of the Society. I did not know him well, but that changed very soon. An IAS officer of the Maharashtra cadre (1951 batch), B.G. Deshmukh was known all over the state for his honesty, no-nonsense attitude and meticulous adherence to rules. I saw these qualities being demonstrated in his BNHS role as well. He had a methodical approach: in the board meetings, he would give everyone an opportunity to express their views – sometimes controversial, sometimes highly opinionated – and with his administrative acumen, he would find a solution that was acceptable to everyone.

An old-fashioned gentleman who came to office immaculately attired in formal wear, B.G. Deshmukh showed us the other side of his personality when he was in the field with us, at the Conservation Education Centre (CEC) in Mumbai, where he would interact with children and BNHS members, enquire about trees and flowers, and ask questions on natural history. He would always see to it that every visitor was comfortable, and nothing would miss his sharp eyes. I admired his persona, which exuded love, care, concern and affection for everyone, including irritating varmints who would pester him for attention. In the initial years of my directorship, I would visit him in his large office at the Bombay House, an impressive neo-classical building, about half a kilometre from Hornbill House. Incidentally, the main office of Tata Sons is also located inside the Bombay House. Evening meetings would invariably lead to tea and pastries, set out in the finest crockery and served in the typical British style.

My respect for B.G. Deshmukh reached its pinnacle when I returned a cheque of nearly a crore to the Tatas, as they were not willing to accept our conditions for taking up studies on the Olive Ridley Turtles (*Lepidochelys olivacea*), which we thought (and rightly so) would be impacted by the Dhamra Port (Odisha) that they were building, close to the famous Rushikiliya turtle nesting site. After a protracted negotiation with the Tatas, when they decided to go ahead with the construction, I decided to return the cheque. B.G. Deshmukh, despite being on the board of several Tata enterprises earlier, never influenced me. When I told him the full story and my apprehensions, he said, 'Rahmani, you take the decision that is good for BNHS and for conservation.' I think BNHS was lucky to have a president like him.

If B.G. Deshmukh was a pillar of BNHS, J.C. Daniel was its very foundation. My association with J.C. goes back to the early 1970s. I do not remember when it was that I first wrote to him, but I do remember the promptness of his replies. In those days we did not have the Internet and used what is now derisively called the 'slow mail', but J.C. Daniel's quick replies would shame even current Internet users. His replies were pithy, to the point and affectionate. What I particularly liked was his long signature, which looked like a long, stretched 'H'.

I later came to know that the width of his signature depended on his mood!

As the curator of BNHS (effectively its CEO), he was my official boss for 12 years, from 1980 to 1991. But to me, he remained 'the boss' even when I was in AMU (from 1991 to May 1997) and more so when I rejoined as director in 1997, though by then he was the organization's honorary secretary. We were truly lucky to have him as a boss – strict with staff, yet caring, J.C. Daniel was devoted to BNHS. Even people who did not like him could not question his devotion to BNHS and wildlife conservation.

J.C. Daniel steered BNHS through a difficult transition period in the 1980s, as it transformed from a small natural history institution to India's premier conservation organization. Strongly resisting the pressure from some EC members to keep it as a 'members-only organization', where members would do the projects (in their spare time, of course!), J.C. Daniel supported the idea of having scientific staff to work on projects full-time. In this struggle, he had the strong support of Dr Sálim Ali, Dr A.N.D. Nanavati, Dr C.V. Kulkarni, Prof. P.V. Bole, and young members such as Cyrus Guzdar, Bittu Sahgal, Dilnavaz Variava and a few others.

Administrative expertise aside, J.C. Daniel's language and natural history observations were commendable. Unfortunately, he did not write as much as his admirers (like me!) wished him to. Besides editing *JBNHS* for over 40 years, and starting the popular quarterly magazine *Hornbill* in 1976, he wrote and edited many books, including *The Book of Indian Reptiles* and *A Century of Natural History and Conservation in Developing Countries*. Not many people know that his paper, 'The Tiger in India: An Enquiry – 1968–69',[24] provided the impetus for the government-initiated Project Tiger launched in 1973.

In the early 2000s, BNHS saw many administrative changes – the director was made the institution's administrative head, the HS's role was minimized and the EC became a policy-making body. With these changes, my responsibilities increased. With fundraising, getting new projects, collaboration with other NGOs, government committee meetings, staff management and administrative tasks now on my plate, I became busier, leaving hardly any time for fieldwork and research. I was working for BNHS 24×7. At tough moments, it was to J.C. Daniel that I turned for advice.

His association with BNHS started in 1950, the year I was born, and continued until he passed away on 23 August 2011. He had been suffering from cancer for some years, and when we heard the news of his hospitalization, I went to meet him along with a few staff members one evening. He shook hands with each of us and said softly, 'Time has come to go'. At 11.00 p.m. that night, I got a phone call from his son that his father had passed on. He will be remembered forever, including by the lizard *Bronchocela danieli*, which was named after him.

The Ali family dominated the landscape of natural history studies and BNHS as well, for almost 50 years after India's Independence. In addition to Dr Sálim Ali, another guru of natural history studies from a young age was Humayun Abdulali (Sálim Ali's cousin). The minutes of the EC meeting of 14 July 2001, held about a month after his passing, succinctly condenses his contributions:

Mr Humayun Abdulali joined the Society as a Life Member in 1931 and was responsible, along with Dr Sálim Ali, for reviving the activities of the Society after our Independence in 1947. He was the Honorary Secretary of the Society during 1950–1962 and was responsible, along with Dr Sálim Ali, for obtaining financial support from the Government of Maharashtra for the maintenance of the collection [of specimens], and from the Government of India for the construction of [the] Hornbill House to house the collections, library and offices of the Society. Mr Abdulali is one of India's foremost conservationists and naturalists, and was responsible for the preservation of the Borivali (Sanjay Gandhi) National Park. He was also involved in formulating the Bombay Wildlife Protection Act, 1951, which formed the basis for the Indian Wildlife Protection Act, 1972. A versatile naturalist, he published over 300 papers and notes on the flora and fauna of the subcontinent, particularly birds. He was elected Emeritus Naturalist of the Society in 1992.

My first encounter with him was during one of my brief stopovers at Hornbill House in the early 1980s, when transiting between Nannaj in Maharashtra and Karera in MP. Whenever in Bombay, I would visit the library and also the BNHS museum. Abdulali was known for his short temper and no-nonsense attitude, so I was quite afraid of him; it was with some timidity that I went over to introduce myself on one such visit, in 1982, when I saw him in the BNHS museum, which was also known as Collection Room as specimens were kept there. He was measuring bird specimens with his assistant and did not approve of my intrusion, so I withdrew silently. After finishing his work, he called me over and asked, 'What do you do?' When I told him that I was studying the GIB, he got excited and asked me many questions. After this initial meeting, I made sure that whenever I was in Bombay, I spent some time with him. When I rejoined BNHS in 1997, I remember going to his Pali Hill residence in Bandra a couple of times to see his papers and books. Age was not on his side but he was alert and active. Seated in his study room in a white *kurta-pyjama* and surrounded by books and papers, he made an imposing figure – a respected family patriarch, which he was.

In 1983 or 1984, Mr Abdulali asked me to show him the bustard, and so we travelled together to Nannaj. I wanted to take the BNHS vehicle, but he insisted on driving himself, as his son Akbar had a new car. In those days, it used to take an entire day to reach Solapur by road. We started early but the Bombay traffic delayed us by three hours. All along the way to Pune, he regaled me with stories of his younger days, his love for motorcycles, cars and fast driving. At about 60 km ahead of Pune, on the Solapur road, there was a small wetland on the right, where he insisted on birdwatching, although evening was approaching. He even forced me to investigate a nest of the White-breasted Kingfisher located in the mud bank near the wetland, to know how many eggs/chicks were there. His curiosity at that age was boundless.

But this energy soon landed us in trouble. By the time we reached Tembhurni town, around 80 km from Solapur, it was already dark and Abdulali looked tired. I requested him quite a few times to allow me to drive, but he would not accede the wheel. Instead, he proceeded to drive like an 18-year-old boy who had just acquired a driving licence! The Pune–Solapur road serves as the main highway for trucks going to South India, so traffic was heavy. A stream of trucks with glaring headlights were coming from the opposite direction, which blinded us momentarily. And bang! In the glare, Abdulali could not see the bullock cart a little ahead of us and hit the cart's large wheel, creating a deep indent on the bonnet. The cart owner disappeared in the darkness, leaving us with a leaking radiator and a stalled car. Abdulali was shaken but calm, more worried about what his son would say. But *my* worry was about how we would reach Solapur! At that time, there were no mobile phones, so there was no way to contact Ranjit Manakadan and Anwar-ul Islam, who were no doubt anxiously waiting for us at the Solapur Circuit House. Finally, I hitched a ride in a Jeep and reached Solapur, where I managed to arrange a vehicle, and also a tow truck for the stalled car. When we reached the accident site, we found Abdulali sitting calmly in his folding chair, sipping hot tea from the famous large thermos that he always carried with him. Many old BNHS members, who had once accompanied Abdulali to the Sanjay Gandhi National Park, still remember his folding chair, the wicker picnic basket full of sandwiches and the large thermos, from which he would offer tea to members. He and his wife were gracious hosts even in the forest!

My next job was to see to it that Abdulali's vehicle was immaculately repaired and painted, as he had to return in a day's time. Thanks to the frequent repairs for my old jalopy, I had developed a kinship with the local mechanic, Abdul Ahmad, who completed the job in time. Meanwhile, we spent the whole day in Nannaj watching bustards, to Abdulali's delight. On the way back, we stopped at the garage in Solapur. When we showed the car to Abdulali, he was surprised and checked the bonnet several times to find a fault. The repair was so well done that Akbar never came to know that his beloved vehicle had undergone a 'trauma'. We all decided to keep mum about the accident. But some days after my return to Bombay, I confessed to J.C. Daniel, who said, 'Rahmani, it was good that you were not driving the vehicle. Otherwise, it would have been the end of your career in BNHS.'

Abdulali was a true naturalist, interested in every species, from spider to tiger. In his honour, a new species of frog (*Nyctibatrachus humayuni*) and the Nicobar Scops Owl (*Otus alius*) were named after him. Abdulali himself described several species, including the Andaman subspecies of Black Baza (*Avecida leuphotes andamanica*), a raptor restricted to the Andaman Islands. He wrote more than 300 scientific papers and short notes, and 60 book reviews, mostly in the *JBNHS*. For his massive contribution to BNHS and bird study, the Bird Room of the BNHS Collections was dedicated to Humayun Abdulali in 2002, in the presence of his family. I feel that if Abdulali had written books and developed a cadre of naturalists, he would have been as famous as his cousin Dr Sálim Ali.

I love Parsees, for they are the most educated, sophisticated and refined community that I have come across. Parsees excel in almost everything, be it business, art, culture, music, literature, science or wildlife, and one is bound to find Parsee experts in any subject. During my time in BNHS, there were always some Parsees on the EC, and they were typically the most forward-looking members, such as Mrs Dilnawaz Variava and

Mrs Pheroza Godrej; they both would always attend all EC meetings despite being terribly busy.

In the 1980s, Mrs Variava developed a business model for BNHS to sell cards and calendars; soon her 'baby' grew into a full-fledged Central Marketing Department, which raised funds for BNHS's wildlife conservation activities. Her association with BNHS was a very long-standing one, going back to 1969, when she had completed her MBA from the Indian Institute of Management, Ahmedabad and became the first paid employee of the newly established branch of WWF-India, which operated out of a single desk in the BNHS library. Later, WWF-India shifted to Delhi and now it has offices in many Indian towns, with headquarters at Lodhi Estate in New Delhi. I remember Mrs Variava always being late for EC meetings, holding haphazardly arranged papers, but her sharp comments and advice helped BNHS in formulating policy decisions.

Mrs Godrej was another bulwark of BNHS, raising funds for the organization. In 2006, she organized a huge art auction at the famous Hotel Taj Intercontinental. It was my first time at an art auction – I was not impressed by modern art (I am lover of classical paintings of the old masters) or by the huge price tag on these paintings, but as long as the funds were for BNHS it was fine with me. Her husband, Jamshyd Godrej, managing director and chairman of Godrej & Boyce, is also an excellent wildlifer. These days many corporates show interest in nature, environment and wildlife (though much of it is merely paying lip service to burnish their public image) but Jamshyd is different. His up-to-date knowledge on wildlife will shame even the so-called 'qualified' wildlife researchers.

The contribution of their company towards conservation

in general, and mangroves conservation in particular, is commendable. Godrej is the largest prime land owner of Mumbai, but that land is mostly covered with mangroves and the company wants to keep it that way. On the Mumbai–Thane road, on one side is located the Godrej company and on the other (towards the sea) lies a long stretch of excellently protected mangroves on land also owned by the company. On many occasions, Godrej has sponsored a Mangrove Walk conducted by BNHS and WWF members. The company even has a full unit, called the Mangrove Cell, staffed with competent officers who help to study and spread awareness about the mangrove ecosystem.

I met Rishad Naoroji in 1981 at the Keoladeo National Park in Bharatpur, Rajasthan. He was one of those people whom one may not like in the beginning. However, if you overlooked his high-strung body language, nervous talk, edgy back-slapping, panicky fiddling with binocs and cameras, tense silences, and frequent and loud clearing of nose, you would unwrap a fine gentleman. For me, his biggest asset was an unquestionable commitment to conservation. We soon became good friends, and would often cycle into Keoladeo. Although he was a scion of the Godrej family and a billionaire, Rishad would not hesitate to eat in a shabby *dhaba*, or sit on the ground with bird trappers to discuss the arrival of migratory birds. I never saw him throwing his weight around by using his family background, or asking any favours for himself. At Bharatpur, Rishad became interested in raptors – a love that resulted in a seminal book, *Birds of Prey of the Indian Subcontinent*, 20 years later. The content and quality of his treatise rivals any similar book in the world.

Rishad was also a member of the EC, but he was not interested in politics, internal bickering and petty machinations, which were the forte of some members. He knew where to focus: BNHS library, books and science. Unfortunately, Rishad developed progressive dementia at the relatively young age of 65–66, and he is not able to recognize anyone anymore. It is sad to accept the fact that my friend, who could recognize an eagle flying 200 m above us, now cannot recognize his old friend sitting beside him. We cannot read his mind but his excellent raptor book and many research papers will continue to enlighten many generations.

Another towering personality I met at BNHS was Dr Erach Bharucha – rather, he was five to six different personalities in one body and that too a lean one. He is one of the best medical practitioners of Pune, where he lives. But his other avatars are: educator, writer, wildlifer, photographer, conservationist, traveller and teacher. He was a member of the BNHS EC, and would come prepared with notes for the meetings and speak eloquently on the subjects he knew. I would take his advice on a range of subjects – from the writing of books to conservation of grasslands, neglected habitats and conservation education.

In 1980, when I met Bittu Sahgal (real name Randhir Sahgal) soon after joining BNHS, he was the 'young turk'. Now four decades on, that epithet still fits! We are of the same age group, more or less, but in outlook and energy he is much younger. Despite the conservation horror stories of the last 50 years, Bittu remains perpetually optimistic, notwithstanding his dire warnings of climate change at the drop of a hat. Bittu started a very successful magazine, *Sanctuary Asia*, in 1981 – one that is still going strong. I consider it the finest wildlife magazine of Asia. He is a natural-born speaker, influencer and leader. I have seen him in sedate government meetings, convincing the politicians and bureaucrats with his arguments, but he is at his best with children. His successful Kids for Tiger conservation awareness programme is testimony to his magic. I have attended a few of these programmes – kids just love him and he loves them.

In addition to those who served on the EC of BNHS, the organization was also able to attract many wonderful supporters and well-wishers. One of these people was Jamshed P. Irani.

I first heard of him sometime in the early 1970s, when I bought a little booklet, *Watching Birds* by Jamal Ara, published by the National Book Trust. Besides the well-written text in simple English, by the first woman ornithologist of Independent India, it had lovely illustrations by one J.P. Irani. Being an amateur artist, I remember copying many of J.P. Irani's illustrations for my fieldwork, I but did not know that I would one day meet the great artist in person.

Irani's association with BNHS goes back to the 1960s, when Dr Sálim Ali was looking for an Indian artist to illustrate birds for his upcoming *Handbook of the Birds of India and Pakistan*, with the famous Sir Dillon Ripley. The main problem with bird illustrations is that unless one has studied the species in the field, one cannot draw correctly from the dried museum specimens. Sometimes, the posture of a live bird is quite different from the stretched up, cotton-filled, eyeless specimens in museum drawers. If drawn purely from these museum specimens, the birds will not be illustrated properly, and what birdwatchers call the 'GISS' (general impression of shape and size) is missing in such illustrations. Wildlife illustration is a domain art, reaching its highest standards in the US and the UK, but has not fully evolved in India. In the 1960s, this gap was filled by J.P. Irani to a certain extent. He became the first bird book illustrator of Independent India, when he did 14 plates for the *Handbook*. Sálim Ali would give him specimens from the BNHS collection, which would eventually be printed on one plate in the book, so that the artist could bring in the subtle differences between them. J.P. Irani recently told me that 'Sálim Ali was a hard taskmaster'.

In 1975, when BNHS was eager to rediscover the Jerdon's Courser (*Rhinoptilus bitorquatus*; the bird that was eventually rediscovered in 1986 by Bharat Bhushan, a BNHS scientist), Sálim Ali asked J.P. Irani to illustrate Jerdon's Courser and Indian Courser (*Cursorius coromandelicus*) for publicity material, which would be distributed in their potential habitat areas. Irani gave the illustrations pro bono to the Society 'as a token of appreciation for the ever-willing help I have been getting from the Society'. Through the years, such kind gestures from supporters have made BNHS stronger.

When J.C. Daniel passed away in 2011, Irani offered to paint a portrait in his memory. We hung the painting in the Mammal Section of the BNHS Museum, on his first death anniversary in 2012. But, not too many people frequented that hall, so it was moved and placed near the painting of Humayun Abdulali, which hangs in the mezzanine floor where bird specimens are kept. As one old member wryly commented later, 'They did not get along during their lives, but at least now they can live in peace side by side.'

I cannot think of my BNHS days without the staff and colleagues who worked with me for decades. There was *chowkidar* Uma Pratap Singh, whose tenure preceded mine! Uma joined BNHS in the 1960s – no one can remember which year. He had the boundless energy of a young man; after a long day at work guarding the gates of the Hornbill House, he would spend time singing loudly, playing with his dog Rani or hanging out with friends outside the gate. As I would typically work late, we became good friends. On a few occasions, he would bring me a dinner of *sabzi* and *roti* that he had cooked, but would get angry if I tried to pay him.

Uma was such a fixture at the Hornbill House that when he went home on annual leave to his village in MP, the House would look empty, Rani would look sad and lost and the staff would also start missing him! As I love dogs, and all other animals, I would look after Rani until Uma returned. Once Uma was back, Rani would ignore me completely, as if telling me, 'My owner is back, I do not need you now!' Uma called it a day in 2014 due to failing health. J.C. Daniel – with whom he had a 50-year association – was also no more. So Uma told me one day, '*Dil oob gaya hai. Ab jane dijiye, Sahab* (I am tired now. Let me go, Sahib).' With great reluctance, I said 'Yes'. We gave him a farewell, on a scale not given to any staff at BNHS, and sent him off loaded him with gifts from the staff.

Another fixture was my driver Parasnath Jaiswal, one of the coolest persons I have ever seen; he joined BNHS in

2002. Paras (as I called him affectionately) was an excellent but timid driver; a few weeks after his appointment in BNHS, I realized that he would follow a larger vehicle, generally a truck, even when there was ample space to overtake the slow-moving vehicle. When I would tell him to overtake, he would do so with great trepidation. If my mind was distracted or if I was reading a newspaper, Paras would be driving relaxedly behind a large vehicle, sometimes for 5–6 km, while the other smaller cars would zip past us. There was clearly a psychological block at work. I started calling him 'Sylvia nana'. A desert warbler, the *Sylvia nana* is a tiny migratory bird of India, and is invariably seen following larger birds, such as wheatears, babblers and bulbuls. After a lot of persuasion and cajoling, it took a year before Paras's psychological roadblock, literally and figuratively, was broken.

Commuting from Chembur to Colaba would take us around two to five hours, depending on the traffic, which was, in the early 2000s, compounded by massive road and flyover construction. For almost 10–12 years, we commuted through huge roadblocks and traffic jams. I would joke to Paras that he was the only staff member with whom I spent most of my time! My friendship with Paras continues post my retirement. Even now, when I visit BNHS, Paras and I often have lunch together.

An organization is like a moving vehicle, whose every wheel, nut, bolt and machine is important. Similarly, every employee in an organization, irrespective of their level, is as important. We need to appreciate the work contributed by people, as long as it's done with dedication and honesty. My analysis is that BNHS became what it did precisely because of this spirit!

12

The BirdLife Partnership

Partnership was key to BNHS, not only in its internal workings but also in its external functioning. One of the longest-standing examples of the latter is the one with BirdLife International (earlier known as International Council of Bird Protection [ICBP]), which spans several decades. During the 1960s and 1970s, Dr Sálim Ali, as the chairman of the ICBP Indian chapter, and BNHS honorary secretary, played a pivotal role in fostering frequent collaborations between the two organizations. The cooperation was largely driven by BNHS's significant focus and achievements in bird and habitat conservation studies. In the 1980s, a joint initiative with the Royal Society for the Protection of Birds (RSPB), funded by the UK government, further solidified this relationship leading to the creation of the CEC in Bombay.

In 1994, the transformation of ICBP into BirdLife International (BLI) brought about a more efficient structure. BirdLife International then evolved into a federation comprising partners representing different countries or regions. It is crucial to note that BirdLife partners are diverse, extending beyond bird-focused groups to include organizations dedicated to various environmental aspects, ranging from scientific research to advocacy. BirdLife International embraces different approaches to conservation, as long as they contribute significantly to bird and habitat protection.

Although BNHS was known all over the world for its work on birds, by the end of 1990 most of its projects were over. Also, by the early 1990s, many good staff had left, with the exception of Vibhu Prakash and S. Balachandran. This created a gap that alarmed many. Many ornithologists had shifted to the newly established SACON. Around this time, SACON pitched itself as a potential BirdLife partner in India, but it did not fit the criteria – it is neither an NGO nor a membership-driven organization.

The wind shifted in 1997 with a change in BNHS's leadership, and my appointment, as I was familiar with BLI. Bombay Natural History Society had also begun realigning its focus towards broader nature conservation, including ornithology. By 1998–99, the goals of BNHS and BLI had become more aligned, and in 1998 the BLI decided to explore the possibility of accepting BNHS as a BirdLife partner from India. BirdLife International has an elaborate process for accepting a new partner, and the admission of newcomers is decided by BLI's regional councils. After working with the regional council for a few years as a partner-designate, with no voting power in any BLI decision, the admission of newcomers is put to a vote at a meeting of BLI's Global Council, held every four years. Bombay Natural History Society was accepted as partner-designate in 1998, and later voted in as a full member in 2008, at the Global Council meeting at Doha, Qatar. I was a member of the Council but excused myself when the voting for BNHS took place.

The Royal Society for the Protection of Birds is BLI's strongest and wealthiest partner. It helps many partner-designates and financially weak partners in biodiversity-rich countries through funds, training and by working side-by-side. In India, the RSPB assisted BNHS for many years and was involved in numerous activities, such as designing project proposals, fundraising, training staff and setting up the Indian Birds Conservation Network (IBCN).

The main aim behind setting up the IBCN was to identify a network of bird-rich sites of international importance in India – called the Important Bird Areas (IBAs) – and to put down actions to conserve them. From the very beginning, the IBCN realized that identification of the IBAs had to be done by involving experts, local NGOs, communities and government. It was primarily a bottom-up consultative approach involving thousands of people. During surveys of potential IBAs, local involvement is crucial so that when sites are finally identified, local NGOs, government bodies and communities already have a stake in the preservation of these sites. The IBCN was basically a people's network.

To collect information for identifying IBAs, BNHS and the RSPB organized state-level workshops for local NGOs and experts. The first workshop was conducted in Rajasthan in 1999, followed by one in West Bengal. Over the next four years, they organized nearly 30 state-level workshops to identify, prioritize and finalize IBAs. During these workshops, I realized that many younger participants, particularly in the Northeast and Central India, did not know how to collect scientific data. To fill this gap, a simple book, *Bird Census Technique*, written by my friends Dr Salim Javed and Dr Rahul Kaul and funded by RSPB–IBCN, was distributed for free to partners during field trainings.

With the support of the IBCN members, scientists and BNHS staff, IBA data started to flow in. We started with suggestions for nearly 3,000 sites – some members had included their 'favourite birding sites' in their state's list – which was cut down to 1,000; but even 1,000 was too much. By the end of 2003, we had narrowed them down to 465 sites. Zafarul Islam, Sunil Laad, Abhijit Malekar, Supriya Jhunjunwala, Farah Ishtiaq, Mohit Kalra and a few others worked long hours to bring out a 1,133-page tome titled *Important Bird Areas of India: Priority Sites for Conservation* (2004).[25] The book had a high-profile launch at the India International Centre, New Delhi, and was released by A. Raja, the then union minister for Environment and Forests. It became one of the most widely quoted books of BNHS.

Later, BLI added biodiversity as a factor in the selection of IBAs (though major emphasis continued to be on birds), which is why designated sites are now known as Important Bird and Biodiversity Areas. I am proud to share that the IBA-IBCN programmes were supervised by me. I trained many young staff members of BNHS, who are now carrying on conservation work in many different organizations. I also got an opportunity to interact with some amazing Indian conservationists during the IBA workshops – it was always a two-way learning process. A popular newsletter, *Mistnet*, was published under the IBA-IBCN programmes, which documented information about IBAs and threatened bird species. I also encouraged IBCN partners to write articles in *Mistnet*, and developed a Writing Training Course for amateur writers-naturalists.

A significant outcome of the BirdLife partnership was the publication of high-quality books, based on the data gathered for the IBCN programme. *Important Bird Areas of India: Priority Sites for Conservation* was our first book; out of 1,000 copies printed, nearly 30 per cent was given to major contributors, university libraries, institutions, Ministry of Environment and Forests (MoEF), state governments, BirdLife partners and corporates. Our aim was to highlight the conservation importance of 465 sites that were 'essential for bird conservation', as Dr Pradipto Ghosh, secretary of MoEF wrote in his 'Foreword'. From the sale proceeds, I established the IBCN-IBA Conservation Fund. Royalties and sale proceeds from all my subsequent books published during my BNHS days were added to this Fund.

It has always appalled me when supervisors and principal investigators put their names first (as a first author) in the publications. All my life, my aim has been to promote youngsters, so I made it an unwritten rule that whoever writes the draft of a paper should be the first author. I followed the same rule for the IBA book. As Zafarul Islam had done most of the draft compilation from the workshop outputs and literature survey in the IBA programme, he became the first author of my most important book. I received a lot of criticism from my colleagues in BNHS for favouring Zafar but I stood by my principles.

While sharing the book with forest officers at the state level, I found that they were not interested in a book covering all of India – their focus was their respective states. The IBA-IBCN programme had created a formidable 'army' of young conservationists in every state, so I started exploring the possibility of bringing out small state-wise books. Over the next three to four years, we brought out IBA books on Sikkim, Jammu and Kashmir, UP and Maharashtra.

During the IBA programme, we also collected and collated a huge amount of data on sites and birds. Zafar and I decided to make use of this data, and went on to write two books, *Existing and Potential Ramsar Sites of India*,[26] and *Duck, Geese and Swans of India*.[27] Both books came out in 2008. In the Ramsar book, we identified 135 sites that met the Ramsar criteria. At that time, India had only 25 Ramsar sites; we now have 85. I am happy to share that at that time 31 sites listed in our book were declared as Ramsar sites by the GoI. No mean achievement!

Back in 2001, BLI had published a two-volume book, *Threatened Birds of Asia*, which was a monumental work. But, there was no book on a similar theme focused just on India; although species accounts from the *Threatened Birds of Asia* were easily available on the BirdLife website, not many bureaucrats, decision-makers and researchers were using it to sieve out India-specific data. So, in 2009 I started planning an India-focused book on this subject, which was funded by my friend Shrenik Baldota, a wildlife photographer and conservationist, and owner of MSPL Limited (an iron ore mining company). For two years, I researched and collated my own information on the 158 threatened and near-threatened birds of India, and in 2012 the book *Threatened Birds of India*[28] was released. Since I believe that bird books should look lovely, I used photographs of every species (except Sillem's Mountain Finch, for which no photograph was available at that time).

From books to building an on-the-ground conservation network, the BirdLife partnership has truly delivered. Sadly, after my retirement, for want of a champion within BNHS, the programme has lost steam, and was neglected and allowed to crumble. The present director, Kishor Rithe, is trying to revive the partnership. I wish him success.

13

The Spirit of BNHS

When I joined as a director, I was quite eager to change the name of BNHS, as it has a parochial label – 'Bombay'. When the Society was established in 1883, many organizations were named after the city of their origin. There was a Sind Natural History Society, though nothing is known about this Society in recent years. There was another, Calcutta Journal of Natural History, which had a short-run and vanished thereafter. The Asiatic Society of Bengal was started in 1784 to encourage Oriental studies; fortunately, it still survives today. There was one more parochial-sounding society, Madras Literary Society that used to publish the *Journal of Literature and Science*, where T.C. Jerdon (the renowned zoologist and botanist) published some papers. Kumaran Sathasivam, an EC member, wrote to me in December 2023, 'This journal was published by the Madras Literary Society. The Society still exists, but publication of the *Madras Journal of Literature and Science* had stopped towards the end of the 19th century itself, after three series of volumes had been brought out.'

I was not the first one to suggest the name change. Back

in 1981, ahead of BNHS's centenary, Dr Sálim Ali wrote in the *Hornbill*,[29]

> Members must now also give serious thought to a suggestion that has been cropping up from time to time, particularly since our Independence, that the Society which has earned national status by its achievements should now 'upgrade' its name to INDIAN NATURAL HISTORY SOCIETY in order to dispel the mistaken notion that it is a parochial association confined to the city of Bombay. It would improve its chances, the pro-changers maintain, of receiving greater financial and other support from the Central Government and private donors if publicly identified as an all-India institution. However, if after a full-scale debate on the pro and cons, if it is finally decided by the general body of members that the name change would in fact be beneficial for the Society, it seems to me that the centenary occasion would be a good starting point.

He further writes, perhaps presciently,

> Apropos this proposition, it has been suggested by an old and well-tried friend of the Society that rather than change an internationally established and well-respected name, and one of deep sentimental attachment for the old-timers, a viable alternative would be for the Society to bifurcate itself, one part continuing as at present and the other to be known as the Indian Institute of Natural History (or words to that effect).

While the first part remained as such, the other part became the Sálim Ali Centre for Ornithology and Natural History. That is of course another story.

As BNHS was an all-India organization, I felt that that status should reflect in its name. I wanted it to be called the 'Indian Conservation Federation' or 'Indian Conservation Union' or something similar. From 2005 to 2010, we often had discussions around the name change but my suggestions were always turned down. I gave the EC numerous examples of organizations and corporates that had changed their names/logos/brand to fit the changing world – one of these was Google, originally called BackRub when it began in 1996. Within two years, the company changed the name. Does anyone remember BackRub now? However, the EC was not convinced until March 2010, when Shiv Sena activists attacked BNHS and threatened to burn the building down if it did not change its name from 'Bombay' to 'Mumbai'. After many meetings, it was – rather reluctantly, I must add – decided that in order to keep the abbreviation 'BNHS', the name would be changed from 'Bombay' to 'Bharat'. An application was filed on 8 October 2013 to the Charity Commissioner and Public Trust Registration Offices under the Bombay Public Trust Act, 1950 – a piece of legislation that ironically retains the name 'Bombay'. Nothing happened for some time. In November 2013, more documents were sent, and yet, nothing happened. In May 2015, a fresh application was submitted. Finally, we were verbally informed that according to new rules, no NGO or trust could add 'India' or 'Bharat' to its name. The matter died out and BNHS is still known as the Bombay Natural History Society.

More lasting than the campaign for a name change was my mission to ensure that, while I was its director, BNHS should lend its name to quality publishing. On the eve of my return to BNHS, a medical doctor, Dr Ashok Kothari, along with another remarkable member of BNHS, the famous marine biologist Dr B.F. Chhapgar brought out a book titled *Sálim Ali's India*, jointly published by BNHS and OUP. It was an instant success, and now it is a collector's item. The introductory article by J.C. Daniel, 'The Unforgettable Sálim Ali', is a delight to read. The second article, 'Ornithology in India: Its Past, Present, and Future' by Sálim Ali (delivered as the Sundar Lal Hora Memorial Lecture in 1971), is valid even after 50 years! This is followed by the 'Mystery Birds of India', which discusses the Mountain Quail (*Ophrysia superciliosa*) and the Pink-headed Duck (*Rhodonessa caryophyllacea*). The rest of the articles and photographs are reproduced from books and journals in BNHS's collection. This book should be read by all young naturalists and conservationists, for it will help them spread the message of conservation through elegant writing, not through boring jargon-filled scientific papers that do not appeal to the general public.

The remarkable team of the two dedicated BNHS EC members, Dr Kothari and Dr Chhapgar, edited three more books: *Treasures of Indian Wildlife* (2005), *Living Jewels from the Indian Jungles* (2009) and *Wildlife of the Himalayas and the Terai Region* (2012), all based on the articles published in *JBNHS*, books in the BNHS library and Kothari's own collections.

Dr Ashok Kothari is a bibliophile, with a vast collection of extremely rare books, manuscripts, old maps and antique pieces of art. Besides being a successful medical practitioner, he is also a tree lover – his clinic is known as 'TreeShade'. His love for books is reflected in his clinic too. As one journalist wrote in 2008, 'Dr Ashok Kothari's clinic resembles a makeshift library, flanked by large cabinets full of books on both the sides. But you won't find medical encyclopaedias and journals here.' During his 50-year career as a medical doctor and conservationist, Dr Kothari has travelled extensively to find books written by the British on the India they knew. He has books from the seventeenth and the eighteenth centuries, ranging from wildlife to Indian heritage.

Not wishing to confine his love of books to himself, Dr Kothari convinced the BNHS Board to allow him to exhibit his own rare books, and some ancient books from the BNHS collection. The ever-supportive Chhatrapati Shivaji Maharaj Vastu Sangrahalaya provided us with glass display cases to keep the books safely. Dr Kothari curated the exhibition, adding his antique objets d'art. His first book exhibition was held in 1992, when Dr Jay Samant was the director. The *Hornbill*[30] reported that, 'The response was excellent and if a fraction of those enraptured visitors who left the exhibition clutching membership forms do sign up, the Society will be much better for it.' With Dr Kothari's help, BNHS has so far organized eight exhibitions; five were held during my tenure. These week-long exhibitions were always a grand success, attracting the general public, book lovers and students from all over Mumbai. For me, the most important value of these exhibitions was the visitation of citizens who would otherwise never visit Hornbill House. Once people enter the hallowed portals of BNHS, their curiosity about nature increases and many become members of the Society.

Besides rare books, Dr Kothari also exhibited British-era paintings and photographs; a striking example was a 1905 photograph of Rani Lakshmibai's palace in Jhansi, from the book *In the Heart of India*; another was a 1923 road map of Bombay. In 2008, on Dr Kothari's suggestion, BNHS started the 'Adopt a Rare Book' scheme. Through the small donations, we were able to collect Rs 5 lakh – all of which was used for the maintenance of the library.

The BNHS library has some very old books, many of which face age-related problems. Every time they are taken out from the cabinets, for reading or consultation, they face some wear and tear. Although BNHS initially received some funding for restoring some of its books, the sources dried up after the pandemic in 2020. What a pity that preservation of literature, art, culture and nature is a low priority now.

The fate of BNHS' old books now depends on the success of another project. In 2011, I read that the state of Maharashtra had archived 6 million documents, and further planned to scan and microfilm over 300 million (3 crore) manuscripts. I discussed the idea with the BNHS librarian Nirmala Barure, who created a plan with the help of Godrej Archives. The idea was to digitize our archival material, with an aim to eventually give access to the public through our website. Until now, all BNHS scientific reports (more than 300), minutes of Annual General Meeting (AGM) and EC meetings since 1921, 2,500 images belonging to Sálim Ali (both his and those taken by him), his correspondence with nearly 250 people and research papers of BNHS scientists have been scanned, digitized and archived. The plan is to also archive important letters written by former and present directors, and office-bearers of BNHS. What was simply an idea in 2011 is now a flourishing activity within BNHS, which will benefit naturalists, historians and archivists all over the world.

Public outreach is a big part of BNHS. For many years, we have been supporting the annual Kala Ghoda Arts Festival, which runs every February. The BNHS auditorium is made available for screenings and workshops. In turn, the festival organizers allow BNHS to run its own stall to sell books and merchandise; proceeds from the sales go towards conservation.

But our greatest outreach success in recent times was the Flamingo Festival, which I started in 2003. T.V. Sowrirajan, a BNHS member and an ardent birder (the kind that prefers binoculars to a camera), first spotted flamingos in Sewri mudflats around 1994. He informed BNHS, and also wrote an article, 'Are there flamingos in Sewri?', in the *Hornbill* that year. Bombay Natural History Society sent a team to Sewri, and after that infrequent visits followed for some years. The flamingos, mostly Lesser (*Phoeniconaias minor*) were found in thousands in the Sewri mudflats and some other areas of Bombay. They arrive by the end of November/December and leave with the onset of monsoon in early June. Peak numbers were seen from April to June, when around 15,000–20,000 were often spotted. During and after the construction of Nhava Sheva bridge, called Atal Setu, flamingos abandoned the area.

In May 2003, we decided to take BNHS members to Sewri for a 'Flamingo Watch' programme; the event was advertised in newspapers and in the BNHS bi-monthly circular, which goes to all members. Keeping in mind the sunlight and tide conditions, Flamingo Watch was organized from 2.00 p.m. onwards, during the low-tide period when the mudflat is exposed. It was for the first time that we were organizing a programme like this. Expecting 500–600 people to turn up, we made arrangements for a minibus to take members from the Sewri Railway Station to the mudflats and back, a distance of about 2 km each way, through an extremely filthy area. But more than 2,500 people, including BNHS members and the general public, turned up. With the response being so overwhelming, even the available vehicles of BNHS were not sufficient to ferry the visitors. Many people walked to reach the mudflats. Their efforts paid off, for once they reached, the sight of nearly 5,000 flamingos feeding on squelchy mudflats, unmindful of people, abandoned ships and litter, flummoxed them. My diary entry for 17 May 2003 reads:

> Due to the sound created by the construction and re-painting of ships, the flamingos were not very close [to us], but our telescopes played magic [...]. The change of expression [on the visitors' faces] after watching these birds through a telescope was lovely to see ... People were really enjoying themselves and discussing with each other how beautiful these birds are. When a flock took wings, [a] sudden silence fell on the watchers, followed by [a] collective 'Wow'!

A hidden natural gem had been discovered by Mumbaikars. The next day, the flamingos of Sewri were splashed all across Mumbai's newspapers. Flamingo Watch, later termed the 'Flamingo Festival', became an annual event of BNHS. Looking at the popularity of the Flamingo Festival, many commercial nature educators started organizing their own events. However, BNHS stuck to one mega annual festival in May or June, purely for nature education. On popular demand, mini-festivals were organized in the winter; the last edition that I organized in 2015 attracted more than 20,000 people.

The event ran for 15 straight years until 2017, when it was abruptly stopped, as the construction of the Nhava Sheva bridge had begun.

But the scent of trouble had been in the air long before. In the early 2000s, the Government of Maharashtra revived an old plan for a long overbridge, from the Sewri mudflats to the Nhava Sheva port. Bombay Natural History Society and many conservation organizations opposed the plan, and suggested shifting the bridge about 500–600 m further south to save the mudflats, but the government would not budge. J. Rego, a member, sent us a handwritten letter on 25 July 2005, which said:

Wake up Mumbaikars! Lest the [illegible words], your city fathers' plan, puts off your avian friends. Your flamingos […] need […] you. Year after year, they nest in the bosom of your city. Will your skyline be better off without them? Sleep no longer, wake up! Knock on the doors of the mandarins of Mantralaya. Ink those petitions. Speak for the flamingos!'

I wrote back to him, 'We have decided to organize an exhibition on "Flamingos" during the Wildlife Week (3rd to 7th October, 2005) at Hornbill House, Mumbai. I hope by that time some flamingos would return to Sewri.' In August 2005, I began planning an official BNHS campaign to save the Sewri mudflats. We decided to bring out a brochure on flamingos, with the catchline *'Thodi jagah hamare liye … aapke dilon mein'* in Devanagari, and 'A little space for us … in your hearts!' in English. We also decided to get a life-size sculpture of a flamingo, which was an instant hit during subsequent festivals.

Noting the popularity of flamingos, the glowing articles in newspapers on the beauty of these birds and BNHS' strident opposition to the location of the Nhava Sheva bridge, the government became alarmed. The Society was not opposed to the bridge but wanted to the government to shift it 500–600 m towards the sea to save the mudflats. We even suggested that the flamingos could become a USP, with a flamingo sanctuary flanking a modern bridge, linking Mumbai to Nhava Sheva and the new airport, but the government did not budge. Instead, to divert attention, they declared a Flamingo Sanctuary in Thane Creek, which was already a naturally protected area. Some parts of Godrej's private mangroves were also included in the sanctuary.

With great difficulty, in 2012, I was able to get a copy of the so-called Environmental Impact Assessment (EIA) report, named the 'Rapid EIA Study Report', created by ARUP and KPMG for Mumbai Metropolitan Region Development Authority (MMRDA). Though the report repeatedly mentioned 'Sewri mudflats', there was no mention of the thousands of flamingos that foraged there. The report also said that the 'Mangroves showed poor diversity, with presence of only Avicennia species'. Among phytoplankton species, the report says, '[P]oor species diversity. Also, species found were to be stress tolerant.'

Interestingly, the Maharashtra State Road Development Corporation Ltd. funded SACON to conduct a study, 'Mumbai Trans-Harbour Link (MTHL) Project: Study of flamingos and migratory birds' in 2006–08, ignoring BNHS, which is located only 5 km from Sewri. Eleven scientists of SACON and two consultants were involved in the study, whose 147-page report was enigmatically silent on the disturbance, or even destruction, of a large part of the mangrove due to the MTHL. Instead, various homilies were given as recommendation, such as '...all the mudflats in Thane Creek are important. [. . .] To improve our knowledge of flamingos and ensure their protection, monitoring population at regional scales at the breeding and non-breeding sites, and international cooperation are crucial.'

With this type of favourable report in hand, BNHS' objections were discarded and work on the MTHL started. The rationale touted by the authorities was that the non-availability of land for an alternative alignment was the main reason for being unable to consider a change in the alignment. Another spurious reason given was that if the alignment was changed, all clearances would have to be obtained afresh, further delaying the project, which had already gone through two failed rounds of tendering.

In 2016, the police stopped birdwatchers from approaching Sewri, ostensibly for security reasons. Sowrirajan managed to visit the area until 2018, but after that he was not allowed either. In the last few months of his visits, he could see flamingos towards the south of the Sewri mudflats (as opposed to till 2018 when they were seen in the whole mudflat), as frenetic Trans-Harbour bridge work was going on. Today, on the southern side, a few fragmented groups are seen in small numbers – the majestic numbers of 15,000+ are gone. The official statement is that the flamingos have shifted to Navi Mumbai and Thane Creek. Meanwhile, the bridge, now called the Atul Setu Nhava Sheva Sea Link, was inaugurated by the PM in January 2024.

Whenever I visit Mumbai and meet BNHS members, many still remember the Flamingo Festivals. Now we can only see construction-related machines, a jazzy bridge and mudflats covered by an invasive mussel called *Mytella strigata*. The pink beauty is gone, but hopefully if the mudflat is restored, the flamingos may return to Sewri.

Conservation is central to the BNHS ethos. It started as an organization of hunter-naturalists, and even after India's Independence in 1947, most of the members were hunters, forest officers, specimen collectors, erstwhile rajas and nawabs, with a sprinkling of civil society members. Their role in India's natural history and incipient conservation activities cannot be underestimated, but now, with nearly 5,000 members, it is a purely conservation-driven organization with a strong research base.

There were several conservation campaigns in Mumbai, in addition to the one for flamingos, and several outside Mumbai. As the director, my first experience of the strength of BNHS members dates back to 1998–99, when a group led by Durgesh Kanikar, John Manjali, Vishwanath and Tejal came to meet me. Goa was under Governor's Rule at that time. Its governor, Lieutenant General Jack Farj Rafale Jacob, PVSM, was interested in wildlife protection. Some BNHS members had formed a 'Bombay Group' of conservationists to protect the forests of Goa, and on the Karnataka side there was the 'Belgaum Group'. Both these groups wanted the thick forests of the Western Ghats on the Goa–Karnataka border to be saved; both got my full backing and BNHS wrote to the Governor on both counts. The 'Bombay Group' was able to convince the Governor to get the Netravali and Medei forests declared as sanctuaries by the Forest Department.

Netravali is contiguous with the Bhagwan Mahaveer and Mollem National Park and the Madei Wildlife Sanctuary to the north, the Cotigao Wildlife Sanctuary to the south and the Dandeli-Anshi Tiger Reserve in Karnataka to the east. All are Important Bird and Biodiversity Areas. Together, they form a substantial part of the protected areas, comprising nearly 1,730 sq. km of the Western Ghats of Goa and Karnataka.

But as soon as the Governor's Rule was over, the local elected representative, with an eye on the vote bank, put pressure on CM Manohar Parrikar to denotify certain areas of Netravali and Madei. Whenever this issue was raised, BNHS would write to the CM to not take that step. We sent a letter on 25 October 1999, and 15 September 2000 and then again on 22 May 2001. In one of those letters we wrote, 'Unfortunately, in spite of appeals and assurances, your government appears to have ignored our advice and is proceeding ahead regardless of the havoc that the denotification of the areas will cause in a vital water catchment locality of Goa.' The CM replied:

> I would like to set the record straight on the issue. The Government of Goa under section 18 of the Wild Life Protection Act has only declared its intention of notifying certain areas as Wild Life Sanctuaries. The position on the ground is quite different in so much as that there are sizeable habitations within these areas proposed to be notified. In addition, the wildlife population numerically in these areas is not significant as compared to the human population, which are also largely dependent historically for survival on agricultural cultivation [in] these areas. Though the sentiments expressed in your letter are well appreciated, going ahead with the notification would

infringe on the basic rights of the people of the area, which as you would agree, no responsive Government would like to infringe upon.

Fortunately, BNHS members' campaign was successful and both the sanctuaries are now well protected. These days, we do not hear about denotification of sanctuaries by local politicians. The Supreme Court ruling that any denotification of an officially declared protected area, even a small part of it, has to be approved by the National Board for Wildlife, has given great strength to the conservation movement in India, and Netravali and Madei are fine examples.

The BNHS's conservation efforts extended overseas as well. In 2007, BLI contacted BNHS to inform us that Tata Chemicals, together with the Tanzania National Development Corporation, was planning a soda ash extraction plant in Lake Natron, famous for millions of Lesser Flamingos. Lake Natron is the only site where birds in the East African population breed successfully, and is therefore an absolutely vital site for three-quarters of the global population of this species.

In August 2007, BLI wrote to Homi Khusrokhan, managing director, Tata Chemicals Ltd, mentioning the following:

> Nesting Lesser Flamingos are extremely vulnerable to predators, both mammal and birds. Their nesting sites at Lake Natron are inaccessible to mammals, but predatory birds are another matter. Lake Natron is set in arid and

hostile surroundings that offer limited food for avian predators: predator populations are thereby low. The flamingo breeding events are too infrequent and irregular to support a permanent population of predators relying on flamingo eggs or chicks for their food. Very important, there is no permanent human settlements anywhere nearby that could support year-round populations of predators and scavenging birds such as Marabou Storks.'

Interestingly, Homi went to become the president of BNHS in 2011. During the Global Council of the BirdLife International in Kenya, which I attended, I was asked to take up the issue with Tatas, so I wrote to B.G. Deshmukh, requesting him to write to Ratan Tata, enclosing the letter that BLI had written to Homi. BirdLife International then commenced a worldwide campaign in November, with BNHS playing an important role in India. Ratan Tata wrote back to B.G. Deshmukh in November that year:

I can assure you that this matter has been given great thought and attention by Tata Chemicals, who have been working under the guidance of the Species Survival Commission of the IUCN, Geneva, since August 2007. A position statement issued by the Managing Director on 31st October 2007, which I enclose, updates original commitments made to the IUCN and another world authority on this project in August 2007. We have always maintained that the plant will be sited at an appropriated distance away from the Ramsar Wetlands and that we will do nothing to threaten the ecology and wildlife of that area. The choice of alternate sites away from the lakeside was initiated about 3 months ago.

He added, 'I am sure that if BirdLife are not aware of this position, then a message from you to this effect would reassure them.' Thanks to the international and national campaign (by BNHS in India), and Tata's commitment to conservation, the project was dropped.

Conservation action, public outreach through festivals and exhibitions – for me, these embodied the public-spirited nature of India's oldest conservation organization. Not all our campaigns were successful, but we think our voice was heard nonetheless. And that's the card that BNHS should never cease to play.

14

125 Years and Counting

Celebrating birthdays is a joyous moment, an occasion we all enjoy during our younger days, but as we grow old and begin shouldering responsibilities, the fun of celebrating a birthday lessens. But in case of an institute or an organization, it is exactly the opposite – the older, the merrier. Any organization that survives for more than 100 years is looked at with awe and respect; this admiration further increases when you add another 25 years to it.

I was lucky to participate in the centenary anniversary functions of BNHS in 1983, and then its 125-year celebrations in 2008. There was one difference though – in the first instance, I was just a senior scientist without too many responsibilities, but for the 125-year celebrations the responsibility fell on me as the BNHS director. In the 25 years since 1983, BNHS had grown into a big organization. There were also numerous other conservation organizations in India, many established by former BNHS employees, including the Wildlife Trust of India (WTI) set up by Vivek Menon. In this period, BNHS's partnership with BLI had also flowered to a global scale. Not many organizations have survived for 125 years, and hence many BirdLife partners were keen to come to India. During this time, I was a Global Council member of BLI, and chairman of the BirdLife Asia Council. We were running massive programmes together, such as the Vulture Conservation Breeding Programme, IBA-IBCN, Green Governance, Bustard/Florican programmes, besides many large projects; so BLI and the RSPB had to be involved.

A key event in the year-long celebration was a seminar that was organized with the help of Indian Institute of Science (IISc), Bangalore (now Bengaluru). I wanted the seminar to discuss out-of-the-box solutions for wildlife conservation in India, and not the routine ideas of creating more protected areas, strengthening legislation, developing animal corridors, reducing poaching and conservation of tigers and elephants. I also wanted to include new emerging threats to wildlife, such as the future impact on wildlife after India becomes a developed nation in another 20–25 years; new tools of conservation such as commercialization of protected areas to protect them (on the pattern of South Africa); new conservation topics, such as conservation breeding of critically endangered species, restoration ecology, conflict management, reconciliation ecology, conservation in urban and semi-urban areas, culling locally abundant species that can harm rare species and fencing wild areas to protect wildlife; and new partnerships for wildlife conservation with corporates, the army and local communities. At the international conference titled, 'Conserving Nature in a Globalizing India', held from 17–19 February 2009 at IISc, Bangalore, we discussed many of the topics listed above.

Unfortunately, two tragedies happened while I was in

the thick of preparations. On 26 November 2008, terrorists attacked Mumbai, killing 175 people and injuring 300. The iconic Taj Mahal Hotel, about 400 m from Hornbill House, was the main target of the attack, leaving us shaken.

I vividly remember the sequence of events on 25–26 November. On the evening of 25 November, after working until 8 p.m., I went home. Around 11.30 p.m., a telephone call from my friend in Hyderabad, Siraj Taher, woke me; he had heard that an attack was unfolding near Hornbill House. A quick call to the watchman, Uma Pratap Singh, reassured me that Hornbill House was secure, as he had locked the gates and that all the late-working researchers were safe inside. I stayed awake, watching the news all night. The next morning I could not reach the office, as there was a curfew in the area. After the terrorists were killed or captured, the situation calmed down, but the mental trauma remained for the next few weeks. Like most of the staff, I was traumatized for many days and could not concentrate on my work.

I had still not recovered from this trauma when early in the morning of 17 January 2009, I got a call from Dr P. Azeez, senior scientist of SACON, telling me that Ravi Sankaran (my *bindaas* co-worker in the Florican Project) had died of a heart attack. I was very close to Ravi, and the news completely devastated me. I remember how the day before, on the evening of 16 January, Ravi had called to invite me to Coimbatore, for a meeting at SACON on 23 January. I had scolded him mildly, as he knew that I was hard-pressed for time due to preparations for the international conference but he somehow convinced me; then to my shock, he was gone the next day. I had never imagined that on the 17th, I would rush to Coimbatore to attend his funeral. I have seen many deaths in my family, but my younger sister's death at the age of 36, and Ravi's death at 46, still brings tears to my eyes, even after all these years. Some wounds just refuse to heal.

Despite the deep wounds caused by Ravi's unexpected death, I submerged myself in preparations for the seminar. That was the only way for me to escape the tragic reality. We had a jam-packed schedule, with a meeting of Asian BirdLife Partners preceding the seminar, as well as an annual meeting of the Vulture Conservation Breeding Programme – all three events were held back-to-back. Fortunately, the seminar went off very well, with some wonderful national and international plenary speakers, and theme-based sessions. Speakers included Prof. R. Sukumar of IISc, Dr Ravi Chellam, Dr Andrew Cunningham of the Zoological Society of London, Dr Paul Morling of the RSPB, historian Ramachandra Guha, B.C. Choudhury of the WII and many more. It was attended by the who's who of the Indian conservation community and extremely interesting papers were read. I was keen that these papers be compiled and brought out in the form of a book published by BNHS or as a special edition of the *JBNHS*. Unfortunately, we couldn't collect the papers from most of the speakers. Not publishing the proceedings of the seminar was one of my biggest failures of life. The only consolation was that we had brought out the papers' abstracts in a booklet form but abstracts have no academic value. My failure still haunts me.

The 125-year celebrations included an exhibition showcasing BNHS's work, a rare book exhibit, an exhibition on 'Medicinal Plants of India' and several book releases, including one of mine – *Duck, Geese and Swans of India*. Events were held in other locations as well. The BNHS-Green Governance Awards were given out at Kolkata's Raj Bhavan in January 2009 by Gopalkrishna Gandhi, the then governor of West Bengal. On that occasion, the Sálim Ali Awards were given to Romulus Whitaker and Dr Ullas Karanth. As a special event, a Bird Migration Study Centre was inaugurated at Point Calimere, Tamil Nadu, in February 2009. Since we wanted to make the 125-year celebrations memorable, another book, *Natural History and the Indian Army*, edited by J.C. Daniel and Lieutenant General Baljit Singh, was released in New Delhi on 5 October, by the chief of the army staff, General Deepak Kapoor.

One of the highlights of the event was the release of a DVD containing 100 volumes of the *JBNHS*. In 2000, Kumaran Sathasivam, a life member (and now GC member), told us about Joseph Joy, a young man settled in the US. A great supporter of BNHS, Joseph had offered to scan all the volumes of *JBNHS*. Well-known British birdwatcher Andrew Robertson found out that a *JBNHS* set (containing the first 70 volumes) was available for sale in Europe. When he told Joseph about this, the latter bought it at once. The set was shipped to Seattle in the US, where Joseph lived at that time. Bombay Natural History Society provided the remaining 30 volumes. Joseph began scanning the volumes in his spare time. When I came to know of Joseph's offer to produce a DVD, I jumped at it. It took more than eight years to complete this (by then, Joseph had moved back to India), but it was completed in time to be released during the 125th year.

Since its release, we have sold hundreds of copies of the DVD, at a nominal cost of Rs 1,000 each. Imagine having 100 volumes of *JBNHS* for just a thousand rupees. It is my dream that all young wildlife researchers go through the pages of *JBNHS*, as I did during my University days, and read the gems that are printed in its venerable pages.

Another publication that celebrated the 125-year anniversary was a special issue of the *Hornbill* on the ongoing projects of BNHS. It comprised 22 articles, with excellent photographs of study subjects, such as corals, elephants, floricans and hornbills. This issue made a huge impact during the seminar, as each participant at the event was given a copy of the *Hornbill*. Until then, many participants who were not members of BNHS had no idea about the scale and spread of the work that was done by BNHS – from marine ecosystems to the Himalayas. The original submissions varied in quality, so it fell on J.C. Daniel and me to rework them. The language varied from 'Hinglish' to 'Tamilish', and so J.C. Daniel worked hard at converting the contents into English.

While the *Hornbill* special issue of October–December 2008 covered the major projects that BNHS had been doing, the last issue of 2009 covered the 125-year celebrations. It also celebrated the unsung heroes of BNHS – the staff, who, along with members, constitute the backbone of the Society. Some staff members remain in the limelight due to the nature of their work, while scientists get recognition through their research papers and books, but there are staff members who remain hidden behind the curtain, although their role in the running of the Society is as important as that of the BNHS scientists. So, in that last issue of 2009, we highlighted the role of almost every staff member of BNHS, and included pictures of them taken at various events during that year. Both these special issues of *Hornbill* are now collector's item for members and libraries. For me, they bring back memories of the golden period of my life in BNHS.

15

Across International Lines

In 2010, I, along with a BNHS colleague, Mohit Kalra, attended the tenth Conference of the Parties (COP-10) of the Convention on Biological Diversity (CBD) held at Nagoya, Japan. It was a great learning experience, as never before had I attended such a large international gathering. We were a part of the BLI delegation; BLI had organized a session on 'Birds and Climate Change', where it declared the results of a large multi-country project, 'Evaluating the Effectiveness of Conservation Site Networks under Climate Change', in which we had collated information on birds from India's Northeast and the IBAs located in that region. Other countries involved in this project were Nepal, Thailand, Vietnam and the UK.

People who have not attended such a large conference (actually, COP is an amalgamation of many conferences, meetings, seminars, discussion groups, exhibitions, specialized events, corporates, governments and NGOs) would not really know what to do there. It requires participation in a few COPs to fully understand its workings, and to make any significant contribution as a participant. Otherwise, you require the help of an experienced participant, who can explain the workings and importance of COPs, as BLI did with us in Nagoya. My friend, Ashish Kothari of Kalpavriksh (an NGO), had attended a few COPs, so I also spent a lot of time with him in Nagoya. He explained to me the intricacies and international politics of the COP. Unfortunately, for many other government and NGO delegates, COP-style meetings are only a free jamboree to a fancy country on someone else's money.

After two years, in 2012, when India was the host of COP-11, India's MoEF invited me to participate in preparatory meetings. For a long time, the venue was not decided; finally, Hyderabad was chosen for its modern facilities and moderate climate. Considering the Japanese way of doing things – meticulous, well-planned, timely, systematic, clean – I was sceptical of India's preparation. How wrong was I! The COP-11 at Hyderabad was one of the best organized COPs of the CBD. It was like a Punjabi *shaadi*: chaotic, disorganized, unruly and tumultuous, till the *baraat* arrives with *band-baja*, then everything falls into place, leaving the *baraati*s overjoyed, both with the arrangements and the festivities! This is exactly what happened in COP-11. Delegates from nearly 150 countries were astounded by the arrangements and hospitality of India. They did not realize how much background work the GoI had to do to make it successful. On the work front, some extremely important decisions were taken for saving a few endangered species. One of the major outcomes of COP-11 was a strong call to parties, partners and other stakeholders to take urgent actions towards achieving the Aichi Biodiversity Targets (decided in COP-10). The other major outcome was the parties' commitment to double international financial flows for biodiversity by 2015, totalling approximately US$30 billion for the biodiversity-rich countries. Declaration and preservation of marine-protected areas was also highlighted, along with the issues of over-harvesting of fish resources. In the area of cooperation with industry, governments were called upon once again to continue their dialogue with industry, and to promote the establishment of national and regional 'business and biodiversity' initiatives and global business partnerships. The parties were invited to strengthen capacities in developing countries, and to support them in designing political instruments and guidelines, which would help their industries to pursue biodiversity-friendly strategies.

The BNHS Board gave me full freedom to prepare for COP-11. Discussions with BirdLife Asia and BLI also helped me and my team. The Board also sanctioned sufficient funds to take a large team to Hyderabad. I wanted to encourage and expose my young team of scientists to COP, as they are the future of BNHS and India.

As a BirdLife partner, BNHS got the opportunity to play 'host' and helped BridLife partners make arrangements locally. We opened a mini-secretariat in the IBA-IBCN department to take care of requests from the BirdLife partners. I sent a team to Hyderabad, in advance, to see to the preparation of our stall (we had a large stall in the NGO section) along with the accommodation and transport requirements. Finally, the time came for the BLI *baraatis* to arrive, who came in small groups and from different countries. Till they left India, BNHS as their 'host', ensured that they did not face any inconvenience. It was a case of perfect BNHS teamwork, and we were able to take care of our BirdLife friends very well.

The eleventh Conference of the Parties was the first of its kind mega-event in India, dedicated to biodiversity in the context of economic development, environmental policies, sustainability and civil society participation. Apart from reviewing the progress achieved since the earlier COPs, including Aichi Targets, this event played a crucial role in chalking out the future course of actions for protecting the dwindling biodiversity against the backdrop of ever-increasing development pressures. Bombay Natural History Society's presence, as a nodal NGO partner as well as a participant, was phenomenal. It successfully coordinated with Indian and foreign NGOs, well before the COP-11 began. During the COP, BNHS participated in the main conference, and also organized seven supplementary sessions on topics as diverse as vulture conservation, marine conservation, nature education, corporate participation and sand mining.

After the success of COP-11, the GoI proposed the setting up of a National Biodiversity Museum at Hyderabad, for which our inputs were requested. Always advocating the importance of the so-called non-glamorous species, I wrote to Mr Hem Pande, additional secretary and nodal person in the MoEF for COP-11, saying:

> I would suggest that the theme for the museum [...] be on local biodiversity, particularly of Hyderabad's environs. The importance of the boulder ecosystem, which is so characteristic of the region, should be highlighted. I would suggest that the exhibits of the museum [...] highlight the so-called lower forms of animals and plants of the region, for example, the Star Tortoise, which is found in the region and [is] under [the] threat of smuggling. Therefore, it should be an icon of the exhibition, and not a tiger or elephant, which are over-emphasized. I

understand that [the] proposed museum is on national biodiversity, however, a flavour on local biodiversity will attract more visitors.

Some progress was made in setting up this museum, but when the government changed in 2014, the priorities changed and soon it fell below the government's radar. A pity, as it would have been a novel experiment in emphasizing the importance of local biodiversity.

The issue of sand mining was our focus when it came to agenda setting for the next COP meeting in 2014. Illegal sand mining had flared up as an important conservation as well as an administrative issue. In July 2013, Sumaira Abdulali (BNHS board member, and founder, Aawaz Foundation) and I wrote a letter to Jayanthi Natarajan, the then minister for environment and forests, to include sand mining in the agenda for COP-12. We wrote, 'It would be very apt for India to take the lead in [controlling] coastal sand mining during this special period when we are the custodians of the Convention of Biodiversity. It would be important to mainstream this issue for inclusion in the next CBD COP-12, scheduled to be held in the Republic of Korea in October 2014.' The idea was accepted – at that time, MoEFCC and the government in general was much more inclusive and accepted the role of NGOs in policy-making.

Slowly, other countries also included illegal sand mining in rivers as their agenda. It was a prescient choice; today, sand mining is one of the most disastrous environmental issues, and 'the New Gold' is the second most exploited natural resource in the world. Experts say that at the current rate of exploitation, we will run out of sand by 2050. When people look back, I hope they will say that BNHS flagged the sand mining issue way back in 2012 by organizing a discussion on the issue during COP-11 and subsequently by getting it included in the agenda of COP-12, making illegal and unsustainable sand mining a conservation issue in many countries. In India, this was one of the key outcomes of that COP, and one that BNHS helped shape.

Bombay Natural History Society was represented in one subsequent COP as well – COP-13 held at Cancum, Mexico, in 2016. Ironically, the event was held in a large hotel, spread over acres, built over a large mangrove area. Imagine an international biodiversity conference being held in a space that was built after destroying mangroves! Unfortunately, after 2016, BNHS was not represented in any future COPs, and in my view that is not a good thing. If we want to become a *vishwaguru* of conservation, I think it is important to attend international meetings, for it would help take forward the India-sponsored agenda.

16

The Power of Partnership

Bombay Natural History Society takes up a campaign only when the Society finds the issue important and we have sufficient reasons to oppose any so-called 'development project'. However, BNHS members are free to take up any campaign, and BNHS supports them based on the merit of the case. One such case was the 'redevelopment' of a famous zoo in Mumbai.

The story begins in 2004, when PETA-India filed a writ petition against the pitiable condition of animals in Veermata Jijabai Bhosale Udyan, Mumbai, on 3 November. The Bombay High Court ordered the constitution of an expert committee, in which the director of BNHS was also included, to prepare a report on the zoo. Later, PETA brought out a photographic report in 2005, which captured the horrors of the zoo. Instead of improving conditions at the zoo, the Brihanmumbai Municipal Corporation (BMC) came up with an ambitious plan to totally revamp it, 'on the patterns of the Singapore Zoo' as one of the zoo officials told me. I think the main idea of the government was to use the court order to take over the zoo, built on prime land in South Mumbai, and convert it into 'a playground' for rich people.

Veermata Jijabai Bhosle Udyan, also popularly known as Rani Bagh, whose original name was Victoria Park, was established in 1861 as a botanical garden, and developed by the Agri Horticultural Society of Western India. Inaugurated on 19 November 1861, Victoria Gardens was the first grand project undertaken by the administration post 1857, when the British Crown assumed direct control over the Indian empire, displacing the East India Company. It predates monuments like the Victoria Terminus (now named Chhatrapati Shivaji Maharaj Terminus), the High Court and the Municipal Corporation of Greater Mumbai. Since, not too many people come to see plants, 15 acres were added in 1890 to establish a proper zoo, purportedly 'to add to the pageantry of nature there, as it were…' in the hope that live attractions would bring in visitors. Soon it became famous as *chiryaghar* (Urdu/Hindi name for zoo). In the 1990s, additional adjacent plots were acquired, bringing the total area of the zoo to 60 acres.

The garden was laid out in the Renaissance Axial Planning design popular with many European nineteenth-century botanical gardens. It is adorned with heritage monuments, such as the Sassoon Clock Tower, the Dr Bhau Daji Lad Museum, the Triumphal Triple Arch Screen, the Woodland Conservatory and beautiful Graeco-Roman–style statues. The botany departments of Mumbai colleges use the botanical garden as a living laboratory for their field studies. It is also one of the most visited zoos in India, admitting nearly 8,000 people every day, a number that touches 40,000 on public holidays. In addition, it is Mumbai's largest green public space and thus a vital green lung.

Recognizing the importance of green heritage, Veermata Jijabai Bhosle Udyan was accorded a Grade II-B heritage status, and the Mumbai Heritage Conservation Committee (MHCC) has categorized the botanical garden as a 'special feature'.

The beauty of the gardens did not, unfortunately, extend to the zoo. Following the case filed by PETA, in May 2005 the government-appointed expert committee inspected the zoo and subsequently issued a report suggesting long-term, short-term and immediate actions, including providing partners to solitary animals, such as the zoo's rhino. In response, the BMC came up with a proposal to 'redevelop' the 60-acre zoo and botanical garden, and set up a so-called international zoo at an astronomical cost of Rs 433 crore. This ambitious plan included Australian, African and Asian sections, entailing the construction of several new animal enclosures, and the laying of new sewage, water and electrification lines that would completely destroy the Grade II-B heritage status botanical garden. The garden has more than 3,200 trees of 226 species, including some old baobabs, making it an oasis in the city's concrete jungle. The proposed new plan had high-tech expensive facilities (3-D theatre, auditorium, aquarium, 100-car park, theme parks, night safari and staff quarters) in an already crowded 60-acre compound. In effect, it would result in the cutting of hundreds of trees and removal of heritage structures.

When some BNHS members read this proposal in the newspapers, they were horrified. One such group, which I named the 'girls' gang', led by Hutokshi Rustomfram and Shubhada Nikharge, came to BNHS for support. I gave them my wholehearted backing, and also informed the President and Honorary Secretary. We needed more information, not just the newspapers reports. The 'girls' gang' swung into action and asked the BMC for a blueprint of the plan. With great difficulty they were able to get it, and they shared it with me. We decided to fight for the survival of more than 3,000 trees, many of which were heading for the axe.

For almost 10 years, the 'girls' gang' fought for the protection of Veermata Jijabai Bhosle Udyan. They had numerous meetings, site visits and letter-exchanges with officialdom of the BMC, which exasperated them sometimes, but they were determined to fight till the end. They even established the 'Save Rani Bagh Botanical Garden Action Foundation'. Bombay Natural History Society gave them their full support and many meetings were held at the Hornbill House. We suggested to the Maharashtra government that they keep Rani Bagh as a botanical garden, as it was originally designed and develop a 'modern zoo' elsewhere, with large cages for animals to roam free. It was a long story of struggle, with a large number of Mumbiakers joining the movement that was backed by BNHS.

Prominent Mumbai citizens, such as Justice B.N. Srikrishna (Retd.), Justice S. Variava (Retd.), B.G. Deshmukh, Bittu Sahgal and Sumaira Abdulali, wrote to the CM of Maharashtra. The 'girl's gang' went to New Delhi to meet Jairam Ramesh, the then minister of forests and environment, who then wrote to the Maharashtra chief minister:

> Knowing the V.J.B Udyan as I do for long, I find a lot of merit in the arguments being put forward by the Save Rani Bagh Botanical Garden Action Foundation. I request you to please have this matter re-examined in the light of the

concerns that have been expressed by such cross-section of citizens of Mumbai city. The next year will mark the 150th anniversary of the Botanical Garden and it would really be a pity if the diversity in the Garden were to be destroyed in the name of zoo expansion.

Even Dr Nigel P. Taylor, the then curator at the world-famous Kew Gardens in the UK, sent a similar letter to the Save Rani Bagh Botanical Garden Action Foundation which was shared with the BMC Chairman.

To highlight the importance of the garden's heritage trees, some of which would be chopped under the makeover plan, the 'girl's gang' brought out a large coffee-table book. It was released at Hornbill House in December 2012 by D.M. Sukthankar, former commissioner, Municipal Corporation of Greater Mumbai, in the presence of Dr Pheroza Godrej, president of Friends of the Trees, to the huge applause of the audience consisting of BNHS members, media, concerned citizens, corporate staff and zoo lovers. There was such a big crowd at BNHS that the auditorium could not accommodate everyone, so a big screen was set up for people who were forced to stand outside.

Under tremendous pressure, the Government of Maharashtra scrapped BMC's plan; instead, it took measures to improve the existing zoo and make it user-friendly; this, in turn, saved all the trees.

The struggle and subsequent success of the Save Rani Bagh Botanical Garden Action Foundation proves the impact of collective and determined action of informed citizens. We are lucky to have members like Hutokshi, Shubhada, Renee Vyas, Katie Bagli and Hutoxi Arethna – all part of the 'girls' gang'. I am proud that I gave every possible support to these passionate members of BNHS who are concerned about plants, animals, the general public and the green lung of Mumbai. More power to them!

What I like about BNHS is that it is always willing to collaborate on conservation issues; this is perhaps due to its basic character of being a member-based organization. The Indian armed forces have a huge area under their control, not only at the border but also in the form of their well-maintained, sprawling cantonments and army depots where munitions are kept. As munitions cannot be concentrated in one or two spots, they need to be dispersed, hence the presence of many large army depots. These areas are also heavily guarded, and consequently have low public footfall, so they become havens for wildlife, particularly birds. Army-firing ranges, such as the one in Khetolai in Jaisalmer, are also excellent wildlife areas. Incidentally, the second-largest breeding population of the critically endangered GIB now survives in the field firing range of the Khetolai–Pokhran area.

Earlier, army officers used to hunt animals, sometimes for target practice, but this does not happen now, except in very rare cases. Actually, the army is now the custodian of wildlife in many areas. I have met many army officers in Rajasthan, Ladakh and Sikkim who are proud of the wildlife found in their jurisdiction. In remote areas, such as Changthang in Ladakh, and in north Sikkim, wildlife research cannot be done without the help of our army. Many times, I have stayed in army camps and enjoyed their famous hospitality. Once they know that you are doing genuine work, army officers will typically go out of their way to help you. For example, one of my students in Kashmir, Mohsin Javid Munshi, is working on the Alpine Musk Deer (*Moschus chryogaster*), for which he had to deploy camera traps, sometimes on the border areas. He is getting a lot of support from the army, despite the fact that the India–Pakistan border is a 'hot' conflict-prone zone.

Bombay Natural History Society used to have an 'Army Cell', which was almost defunct when I joined in 1997. It was an old BNHS tradition to invite chiefs of army, navy and air force to the Hornbill House, and many of them visited us. Earlier Dr Sálim Ali, and later other presidents, would take the distinguished guests to the BNHS museum and library, and show them our rare specimen collection and rare books. I revived the Army Cell with the support of B.G. Deshmukh, J.C. Daniel, Bittu Sahgal and Deepak Apte. Bittu brought in Major Phatak, a retired and cultured army officer with great interest in wildlife. Despite his retirement from the army, Major Phatak had vast connections with army officers, and the BNHS-Army engagement recommenced. This was not a PR exercise but a genuine desire to collaborate with our armed forces for wildlife conservation. For example, General Bipin Rawat, chief of defence staff, visited BNHS in November 2021, one month before his tragic accident in a helicopter crash on 8 December 2021. I attended the BNHS function and was amazed to hear him speak about Indian wildlife and conservation problems. His knowledge of wildlife would shame even a wildlife scientist – he was so up to date. His tragic death saddened everyone, including me, as I had interacted with him regarding the bustards in the field firing range of Jaisalmer, and he had promised all help to BNHS in its bustard programmes.

The Power of Partnership

As part of the conservation awareness programmes, BNHS and the armed forces have jointly organized seminars in stations such as Kargil, Leh, Kolkata, Jaipur, Jodhpur, Bhopal, Jammu, Udhampur, Ambala, Amritsar, Srinagar, Tezpur and Nagrota, involving the military stations, naval stations and territorial army battalions. For one of these in July 2016, the Head of the Army's Southern Command presided over a seminar on the 'Protection of Endangered Species – Great Indian Bustard (GIB)' at the Jodhpur Military Station.

Another joint initiative between the armed forces and BNHS was the 'Green Governance' programme. Started in 2004, the programme's aim was 'biodiversity conservation', and was funded with support from the financial institution ICICI. It was led by Deepak Apte, and on his suggestion we constituted the 'Green Governance Award' for the armed forces. The Green Governance Awards, given each year until 2017, were presented to different branches of the armed forces. These were high-profile events, with the then PM Dr Manmohan Singh presiding over the 2005 edition. I had the privilege to deliver the 'Vote of Thanks', which I did after many rehearsals. Anyone can be nervous in front of a distinguished person like Dr Manmohan Singh. I personally think that Dr Singh, with his erudition, calm composure and foresightedness, was one of the finest PMs of India. I am his life-long admirer.

I sincerely hope BNHS will revive the Green Governance Award, and expand the Army Cell to cover the whole country. Defenders of nature and defenders of borders have the same aim – to make our country strong in every field.

Book publication is another area in which BNHS used collaboration to great effect. The Society has a good publication record, going back more than a century. The first book published by BNHS was *The Indian Ducks and their Allies*, by E.C. Stuart Baker in 1908. The book went through many editions, but unfortunately most young ornithologists do not know about this wonderful book. *The Book of Indian Birds* by Sálim Ali, first published in 1941, is now in its fourteenth edition. It is a 'bestseller' for BNHS even now. The Society's second most popular book is *The Book of Indian Animals*, by Stanley Henry Prater, first published in 1948. It has undergone many editions and seen many competitors, but it is still referred to by scientists and researchers.

As director, I promoted the publication of books by staff and members. The Society's efficient Publication Department mainly looks after the publication of the *Journal* (started in 1896) and *Hornbill* (1976), and later *Buceros* (1996) and *Mistnet* (1999), but regular book publication is a different ball game. Moreover, as books are mostly for the general public, not necessarily for members or subscribers, we have to see to their distribution and saleability. Fortunately, BNHS had a collaboration with OUP, New Delhi, under which OUP used to distribute and sell our books. Our agreement with OUP – they retained 45 per cent of the books' MRP – looks unfair at first glance but for BNHS this arrangement was just fine. I know that it is relatively easy to publish a book but quite difficult to distribute it worldwide (I speak of the time before Amazon and Flipkart). Oxford University Press with offices in 25 cities/countries, and agreements with all major bookshops/retailers worldwide can sell hundreds of copies of BNHS books, which we could not have sold anyway from

Hornbill House. This arrangement, started in the early 1980s, flourished in the 1990s, 2000s and 2010s. During my tenure from May 1997 to July 2015, 60 books were published by BNHS, including many reprints. Even if the reprinted books are removed from the list, around 40–45 new titles were published in 18 years, some by me, others by staff and BNHS members. I am proud of this achievement. Unfortunately, the arrangement with OUP ended in 2020. There are many reasons behind this, though it will not be appropriate for me to discuss those, as I was not in the decision-making process at that time.

Publication of books for the sake of it was not really the aim of this initiative – it was also for conservation action and awareness. Therefore, we made sure that a few copies of our books went to decision-makers. In as many books as possible, we managed to get messages from CMs or forest ministers. We also tried to get the books released by prominent personalities, and made sure that the right kind of publicity was done – the aim again was to send the message of conservation across by every possible means.

Some books, such as Isaac Kehimkar's *Common Indian Wild Flowers* and *The Book of Indian Butterflies*, and Deepak Apte's *The Book of Indian Shells*, aimed to fill gaps in current knowledge. They also sold well. BNHS was specially commissioned to bring out a few books such as *Fauna and Flora of Raj Bhavan* by Naresh Chaturvedi, and *Birds of the Rashtrapati Bhavan* by Dr G. Maheswaran – these were prestigious assignments for BNHS.

One of the best such commissioned books was a publication supported by GAIL (India). It documented local customs, traditional system of medicine, food preparation, irrigation, water storage, architecture and handicraft. It took the author

Sunjoy Monga, a prolific writer and photographer, more than a year to write, as his research took him to many areas of the country to photograph the traditional ways of life, crop management, watering systems, dresses, dances and artefacts. Titled *Conservation: Lessons from Traditions*, the book came out in 2010; it featured many interesting nuggets of information, as the following chapter names suggest – 'Learning from the Ant', 'The Sacred Lotus', 'Wisdom of the Backyard' and 'Smart Energy'.

GAIL picked up all the 1,000 copies, due to our agreement with them. If they had allowed us, we could have sold some copies to our members and public at large, as this wonderful book deserved a wider circulation.

I published 24 books (including a few books translated into other languages) starting with *Birds of Mathura Refinery* (first as a booklet in 1998, and then as a coffee-table book in 2001), and ending with *Threatened Birds and Biodiversity Areas of India* in 2016 (there was a limited print edition, but the book is available on the BNHS and BirdLife websites for wider use).

One of the most interesting books that we got to publish was a birding memoir by a Canadian named Michael Spencer. Although a lot of his birding was done outside India, he put together an interesting manuscript, which shows the importance of keeping field notes. The delightful book – *The Accidental Birdwatcher* – also had an evocative cover by a Canadian artist named Jerome Couelle. Another book that I am quite proud of is Dr A.J.T. Johnsingh's *Walking in the Western Ghats*, which went for reprint. His other book, *On Jim Corbett's Trail*, also did well and BNHS sold hundreds of copies.

Katie Bagli, an active life member and a well-wisher of BNHS, is a prolific writer who has released around 52 books. So, I was happy when in 2012, she told me that she wanted to write a book for children to make them aware of natural history. I promised to give her all our support, and thereafter she would spend time in the BNHS library researching for her books. In 2021, BNHS published *The Flight of the Pink-headed Ducks and Other Stories*; she has now written another book, *Fascinating Fungi*, which BNHS is planning to publish.

As Alice Hoffman, the American writer, said, 'Books may well be the only true magic.' Since my childhood, I have been in the grip of this 'magic', and make sure that I read at least two books every month. Unfortunately, the habit of book reading is declining in youngsters and they read only what is relevant to their work, and that too mostly on Google, often without checking its authenticity. I have seen experts in BNHS in the 1980s and 1990s, who were known for their depth of knowledge, not only in natural history but in other subjects as well. For example, J.C. Daniel was interested in English literature and Dr Sálim Ali always kept English or Persian books near his pillows for night reading. Now, youngsters fall asleep holding their mobile phone. I doubt if any magic emanates from a backlit screen.

17

The Legendary Ali Hussain

I first met Ali Hussain in October 1980, when I moved from Point Calimere to Bhatarpur to work on the Bird Migration Project. Ali Hussain was among the 10–12 trappers belonging to the Mirshikari (who were Muslims) and Sahni (who were Hindus) groups, and was employed by BNHS to trap birds for ringing. Ali Hussain was the most talkative, active and domineering among them. You could not ignore him, nor did he want to be ignored; this was a man who was proud of his traditional skills in catching birds with great expertise, with almost zero injury to the trapped birds. His knowledge of birds, their behaviours, migrations and scientific names was amazing, and he loved to show it off. Like the others, I too was impressed by his personality – he had a tall, slim, erect carriage, was usually clad in white *kurta* and white *dhoti*, and sported a two to three day stubble on his sunken cheeks. He had a broad smile, which exposed pure white teeth (cleaned every day by chewing a *miswak* of *Salvadora persica* sticks). Ali Hussain was the first to respond when the trappers were called; instead of the traditional Muslim *salaam*, he would greet everyone with

(L to R) Asad Rahmani, S. Subramanya, Dr Sálim Ali and S. Haque in Keoladeo NP, in 1980
Photo courtesy: BNHS

Indira Gandhi releasing the pictorial guide in 1983 during the centenary celebrations
Photo courtesy: BNHS

Sálim Ali, PM Lad (centre) with Usha Lachungpa and Ali Hussain in Sailana, Madhya Pradesh, in 1984
Photo by Asad Rahmani

Watching birds, Madhya Pradesh, 1986
Photo by Ravi Sankaran

Bustard conservation campaign in the Thar Desert in 1998
Photo courtesy: Manoj Kulshreshtha

Asad Rahmani in Lakshadweep, 2007
Photo by Deepak Apte

Tern nesting on Pitti Island
Photo by Asad Rahmani

Discussing the birdlife of Narcondam Island with Shirish Manchi
Photo by Dhritiman Mukherjee

Discussing the vulture crises in Delhi with Chris Bowden of RSPB (centre), Vibhu Prakash (black jersey), Nita Shah (back to camera) and others in 2011

Photo courtesy: SAVE

Inside a cave in Chalis-eak, Andaman in 2013 with Akshaya Mane and assistants

Photo courtesy: Asad Rahmani

Bengal Florican fitted with a satellite tracking device and ready for release in Pilibhet Tiger Reserve, 2014

Photo by Rohit Jha

Examining an electrocuted Sarus with Bridesh Kumar, 2017

Photo by Fazlur Rahman

Asad Rahmani with Sujit Narwade on a Thar Desert survey in 2014
Photo by Dhritiman Mukherjee

Counting fallen seeds in a harvested crop field with students, 2019
Photo by Prakash Mehta

Asad Rahmani (extreme left) with the Governing Board of The Corbett Foundation in a restored habitat within the buffer zone of Bandavgarh NP, 2023

Photo courtesy: TCF

Asad Rahmani with a villager in the Thar desert during a bustard conservation campaign

Photo by Dhritiman Mukherjee

a namaste, as he lived in Manjhaul village, Begusarai district, Bihar, in a mixed colony, reflecting the syncretic culture that is still prevalent in villages.

Ali Hussain was 'discovered' by Dr Sálim Ali and his team in 1964, when a BNHS team went to Bihar to ring birds under the Southeast Asia Treaty Organization-sponsored multi-country programme to study bird migration in Asia. P.V. George, a research assistant, was sent to Bihar to explore the possibility of bird trapping for research; BNHS was earlier ringing birds episodically, but in other parts of India. The year 1959 was significant for Indian ornithology and the study of bird migration, as this was when the first-ever organized scheme for bird ringing and migration study in the subcontinent was taken up by BNHS. During this programme, which ran from 1959 to 1973, hundreds of thousands of birds were ringed. Ali and his team were a part of the project in North India.

When the Bird Migration Project funded by the USFWS commenced in the 1980s, BNHS started two major field stations: one in Keoladeo National Park, Rajasthan, and another in Point Calimere, Tamil Nadu. Later, many small field stations were set up at Harike, Karera, Nalsarovar, and Chilika. Ali Hussain and a few other Mirshikaris were employed by BNHS for the project, which also included another group of Sahni trappers. The two groups would catch the birds at night, and early in the morning we would ring, take measurements and release the birds. This was mainly for waterbirds; for terrestrial birds, we used mist nets in the drier parts of the sanctuary. In this, we were assisted by young field assistants from the nearby villages. I am happy to say that some of them, such as Randheera and Brijendra, are now top-class bird guides, known all over the world.

Born in 1948, Ali Hussain and his wife Searun Khatoon had four sons and two daughters. Their eldest was Mehboob Alam, about whom I have already written, while another son Qasim Hussain was employed by BNHS in its Vulture Conservation Breeding Programme; two other brothers, Sikandar and Amjad Hussain, also work for BNHS. Other members of Ali Hussain's family, including several nephews, all grew up to be professional bird trappers, whose services are regularly sought after by researchers.

Once the Bird Migration Project in Keoladeo was stopped by Kailash Sankhala in 1984, the trappers were sent back home to Begusarai, where some became fishermen. Ali Hussain's family started making fans from the fronds of palm trees, which grow profusely in the region. However, whenever BNHS required their services, the trappers duly answered our call. In 1985, when I got permission to colour-band four bustards in Karera, I called Ali Hussain. He came with his son, Mehboob Alam, a gawky young man at that time. As GIB was a Schedule-I species, I cautioned Ali Hussain that he had to be extremely careful, as we were dealing with a rare species – zero injury and zero death was the goal. Ali Hussain, as was his nature, studied the bird's behaviour before putting up his nooses. I baited the bustards by spreading wheat, particularly targeting the displaying males. Every day before daybreak, Ali and Mehboob would be dropped a kilometre away from the targeted area, to put up nooses in pitch dark; a long wait of six to seven hours would follow, until it became hot. Noose traps are mostly used for terrestrial birds or those birds that spend a lot of time on the ground foraging, such as cranes, storks, bustards and waders. Slip nooses made of fine monofilament nylon or 'invisible thread' are stuck securely on the ground and the bird is caught when it walks into the noose area. When the bird tries to escape, the noose is tightened around the leg. Nooses are usually placed along favoured feeding, roosting or nesting sites. It is a well-known animal trapping method, perfected by Ali Hussain, to trap targeted species. It took us nearly a month to catch two males and two females successfully. I named the dominant male, Ashok, and the other one, Sultan. The females were named Rani and Laxmi, but they were hardly ever seen after their release. The males were seen for many years in their respective territories.

Dr M.K. Ranjitsinhji, who came to photograph the bustard, used to sit in my hide, installed near Ashok's display spot. Ashok's yellow leg ring was clearly visible in each photograph. After taking sufficient pictures of Ashok, he wanted to take some pictures of the other untagged birds, which did not have the colour ring, as any good photographer would. Other birds (without the colour rings) came to partake the bait, but Ashok would not allow them. After four hours of sitting in the hide, Dr Ranjitsinhji finally came out, and while returning to the guest house, he irately told me, '*Aap ka yeh kambakht Ashok kisi aur ko paas aane nahi deta* (Your damn Ashok does not allow any other bird to come nearby).'

Coming back to the bustard-catching operation, the nooses were put up early morning and late afternoon – a cooler part of the day during summer. We (Bharat Bhushan, Ali Hussain, Mehboob Alam and I) spent long hours in the small 'gypsy' hut. To while away the long summer afternoons, we would talk about birds and other subjects. Once, I saw a hoopoe and told Mehboob Alam, a young apprentice at that time, that it was one of the most widely distributed birds in the world. Ali Hussain interjected, '*Sahab, do types key hoopoe hote hain* (There are two types of hoopoes, not one).' Taken aback, I said only one type (species) of hoopoe exists in the whole world. Ali said, '*Mai sharth lagata hoon. Book nikaliyae* (I bet. Take out the book).' I took out the *Handbook of the Birds of India and Pakistan* and found that there were indeed two subspecies of hoopoe – a migratory one that is slightly bigger, with subterminal white spots on the crest, and a smaller resident one, lacking white spots on the crest. I asked Ali Hussain, '*Aap ko kaise malum hai yeh, kisi kitab me phada tha* (How do you know this? Did you read it in a book)?' Ali replied matter-of-factly, '*Hum sab jante hai chiryoon ke bare mein, Sahab* (I know everything about birds).' This was typical Ali Hussain!

Anyone who has worked with Ali will appreciate his knowledge of birds, gained during the bird-trapping days, and later while reading books on birds. During bird trapping field trips, when he isn't talking to a rapt audience about his experiences, he reads books or repairs his bird nets. He also has a tendency to suddenly ask an awkward question to a young researcher, or to comment on his/her poor knowledge of birds, much to the consternation of the latter. But considering his value to science, these are minor irritants. At Sailana, during my studies on the Lesser Florican with Usha Lachungpa and Ravi Sankaran, we stayed together in a derelict, leaky guest house, where I had an opportunity to learn about the habits of birds from Ali Hussain. Such earthy information on nesting behaviour, selection of food in different seasons by birds, bird folklore, how a particular bird should be cooked, etc., is not present in any ornithology book and comes from experience and tradition. Ali Hussain's father and grandfather were also bird trappers, so they left a legacy of traditional knowledge, which Ali further enhanced by trapping and observing birds. Once he came under the influence of Dr Sálim Ali and BNHS, his commercial trapping of birds for sale/food ended, but he continued to enrich his knowledge on birds. I think he is the finest bird trapper-cum-naturalist of the Indian subcontinent.

Impressed by his trapping skills – which encompassed nearly 100 methods of trapping, depending on the species and habitats – the USFWS sponsored a one-month trip to the US in June 1998, so that Ali could teach them his techniques. His son, Mehboob Alam, accompanied him, and Mini Nagendra of the USFWS looked after them very well. Ali enthralled American wildlife officials by demonstrating various trapping skills, particularly the clap-trap and noose-trap techniques. His most famous traditional clap-trap technique was even mentioned in a research paper, 'Use of Clap Traps in Capturing Non-migratory Whooping Cranes in Florida', which was published in the *Proceedings of the North American Crane Workshop*.[31] He had demonstrated the method while catching the Sandhill Cranes (*Grus canadensis pulla*) in the Mississippi Sandhill Crane National Wildlife Refuge. The clap trap proved to be an important tool during this field trip. One of its most appealing features was the clap trap's ability to safely catch multiple cranes at once. I was told that he was able to catch 10 per cent of the Mississippi Sandhill Crane population. The traps were fairly easy to build and inexpensive, so if a net had to be cut to take the trapped bird out (to reduce handling time), it was not a great loss. Later, it became a routine capture technique for catching the non-migratory population of the Whooping Crane (*Grus americana*). As the Whooping Crane is a critically endangered species, with 600 odd surviving in the world (as per 2022 data), catching them requires a zero-injury strategy because every individual is important. Ali Hussain's technique was perfect, though no one can guarantee zero risk. In effect, Ali Hussain and Mehboob Alam's month-long US tour, including Hawaii, was very successful, despite the language barrier, and one he still brags about.

Sometime in 1998, the Films Division of the GoI approached BNHS, as they wanted to make a documentary on Ali Hussain. I called him to Mumbai and took him to the Peddar Road office of the Films Division, where some initial studio filming was done. Later some shots were filmed in the field, and some clips were given by BNHS. The result was a documentary called *Birdman Ali Hussain*. I was told by the producer that the documentary was translated into 18 languages and shown all over India, sometimes even in the theatres, before the main movie began. It is still available on YouTube.

Even today, Ali Hussain and his sons and relatives remain much in demand to catch birds for researchers. Although I do not have any 'power' over his activities in any capacity, people ask me to send for Ali Hussain. He too tells them to first get permission from 'Rahmani Sahab' and only then would he agree to come over! In a way it is useful because when researchers make a request for his services, I take a guarantee from them that Ali and his team will be treated properly, and not like labourers. His appearance gives him a typical rustic look but BNHS treats him with great respect; he can now communicate effectively with even the highest officials and feel comfortable in the plushest of accommodations. I have always treated him as my equal and provided him with the same facilities that were accorded to me. Many times, when in the field, for example in Sailana, we stayed in the same room. Therefore, whenever any organization needs his service, I first ensure that he is paid well and treated with respect and equality. So, people are also afraid that if they do not treat him well he may report to me, which he does quite often. Researchers like Dr Devesh Gadhvi of TCF, and Mohan Ram, DFO, Gir, were extremely courteous to Ali when he was required to catch raptors in 2023. The same was the case with the researchers of WWF-India. In all these instances, the Ali Hussain aura and reputation played a major role, not me!

Ali was a valued BNHS resource person. J.C. Daniel once wrote about him in a letter, in 1997, 'At the personal level, I have known Shri Ali Hussain from the time he started working with the Society in 1965 and at the end of 32 years, I still remain amazed at his vast knowledge of the habits of birds. He has remained a man of heart-warming humility and an example of the best Indian human values.'

In 2001, Ali Hussain wrote to me stating that he had been working with the Society as a bird trapper for more than 30 years, and that during that period, his services had been utilized for several projects. He further mentioned, 'I have been instrumental in training many of the young scientists of BNHS.' This was true, as I myself learnt many aspects of bird behaviour and bird identification markers from him. I forwarded the letter to J.C. Daniel; here I must give credit to J.C. Daniel for convincing the EC to give Ali Hussain a monthly 'pension' of Rs 5,000.

Despite his advancing age, Ali is still in demand for bird trapping, not only at BNHS but also at many other organizations, such as TCF, WII, WWF-India, Sher-e-Kashmir University of Agricultural Sciences and Technology (SKUAST) and forest departments. His other major contribution has been that of training several relatives to trap birds for research purposes. They are not as good – or as talkative – as the 'Guru', but nonetheless they are useful for bird movement studies.

18

The Vulture Crisis

'*Ab toh gidh bhee kum ho rahe hain*', an innocuous statement by my former animal-keeper Ram Gopal (who took care of rescued animals at the Centre of Wildlife in AMU) alerted me, and perhaps resulted in the biggest bird rescue project in India. It was sometime at the end of 1996, when I was working as the Centre's chairman, that I heard Gopal talking to his colleagues about the decline of the vulture populations. I at once called him into my room and asked him about it. He showed me a small news item in the Hindi newspaper *Amar Ujala*, published from Agra. The article was on Meerut, a fast-growing town in western UP, about 150 km from Aligarh. After reading the news, I kept wondering why it was published, and, if true, what exactly was happening to the vultures, which were very common in Aligarh and other areas. Nevertheless, I started keeping a watch on vultures, which still filled the skies over Aligarh.

Since joining AMU in late 1991, I used to take students to Patna Jheel, an excellent wetland in the Jalesar *tehsil* of Etah district. On the way was a railway crossing, just after

Sikandra Rao, a crowded and dusty town of UP. On the other side of the railway crossing was a carcass dumping ground, where we usually saw hundreds of vultures, mainly Oriental White-backed (*Gyps bengalensis*), Long-billed (*Gyps indicus*) and Egyptian (*Neophron percnopterus*). They were present at that ground all the time, thanks to the constant supply of cattle carcasses, either jostling with stray dogs to get at the meat, or sitting on old trees after a heavy meal or flying around. A cursory glance was all they got from us until then. A few days after seeing the news in *Amar Ujala*, when I was on my way to Patna Jheel, I stopped at the Sikandra Rao carcass dump to count the vultures. They certainly were lower in number now, only a few dozens, not the hundreds I used to see earlier. After a few months, I had to drive to Dehradun, and en route I stopped at the Meerut carcass dump. I checked with the locals there, and they all said that the *gidhs* had decreased in number. One incredulous villager said, 'Sahib, Americans have taken them away!' On a subsequent visit to the WWF-India office in New Delhi, I was told by Parikshit Goutam (a former student of AMU and then head of the Wetlands Division of WWF-India) that vultures, once common, had stopped breeding on the tall Neem trees along Delhi's avenues, and fewer and fewer were being seen in their favoured haunt, Lodhi Garden.

A visit to Bharatpur further confirmed this decline; Vibhu Prakash, who by 1996 had already been working for 15 years in Bharatpur and knew raptors like no one else, also confirmed the trend. His PhD thesis was on the 'General Ecology of Raptors in Keoladeo National Park' and so, he had excellent census data on all species of raptors, including vultures. A quick comparison of data from 1983–86 with 1997–98 showed a shocking decline of the population of vultures. There was something really going wrong.

In May 1997, I joined BNHS as its director and asked Vibhu to look into this problem more seriously. Was the problem restricted to Bharatpur–Aligarh–Meerut – the so-called cow belt – or more widespread? In 1993–94, I had done four extensive surveys in the Thar Desert, each consisting of not less than three weeks, mainly looking for the Stoliczka's Bushchat (now named White-browed Bushchat) and studying the wildlife in general. During the surveys, my students and I had noted encounter rates of important birds and mammals across a total distance of more than 5,000 km. The vultures were so common then that I never bothered to take detailed notes about their numbers. In fact, just the previous year, in 1996, I had written a short note titled 'Status of Vultures in the Thar Desert of India' in a newsletter called *Vulture News*,[32] which was published from South Africa. At that time, I had no inkling that something was wrong on the 'vulture front', at least in India, although Nigel Collar of BLI had suggested including them in the list of threatened species, mainly due to their decline in Southeast Asia. But why had we been blindsided by the sudden vulture decline? The problem was that the *Gyps* species of vultures, particularly White-backed and Long-billed, were so abundant in North India up to the mid-1990s that even if 50 per cent disappeared, we would still see thousands of them. Secondly, no one in India was monitoring them systematically to notice their declining trend.

As the vulture crisis loomed, the data collected during BNHS studies gained importance. Vibhu Prakash joined BNHS in 1981 in the Keoladeo Ecology Project, and after completing his PhD on raptors of Keoladeo, and many projects on raptors, Vibhu did an extensive study of raptors all over the country, under the project 'Ecology and Behaviour of Resident Raptors with Special Reference to Endangered Species' from 1990 to 1993, funded by the USFWS. He thus had the data for comparison. Still, when his final report was published in 1995, I remember a respected zoologist telling me derisively, 'What is this? Driving and counting raptors from the road. Is this science?' Eventually, Vibhu's simple roadside encounter rate data became a key benchmark to compare the vulture numbers between 1990–93 and 2000–03, and became our biggest tool to prove to the government that the vultures had declined all over India, and not just in a few select places.

But we still didn't know what was causing the decline. In 1998, I organized a small meeting with the limited funds BNHS had. At this meeting attended by BNHS, WWF-India, WII, SACON and vets, various theories were given for their decline, and concerns were expressed. But at the end of that day, we all agreed that we needed more scientific data to find out what was killing vultures. Meanwhile, I wrote a short article in the *Newsletter for Birdwatchers*,[33] published from Bangalore; perhaps the first article, albeit in a newsletter, about the decline of vultures in India. By this time, the Internet revolution had begun to unfurl in India. I remember how Bittu Sahgal, the editor of *Sanctuary* magazine and a good friend, first showed me how to use the Internet in early 1998 at his house. Later, I got an Internet connection installed at my office in BNHS. One of my first emails was about vultures, titled 'Vulture Alert'. The email resulted in interesting discussions that went on for many years. Some of the theories, given in these emails as reasons for the vulture's decline, were downright absurd, while some other comments were illogical and unscientific, questioning our competence, but I will leave those details out lest I ruffle too many feathers.

Based on his 15 years of excellent field study at Bharatpur, Vibhu wrote a scientific paper, 'Status of Vultures in Keoladeo National Park, Bharatpur, India, with Special Reference to Population Crash in Gyps Species',[34] that was published on priority in *JBNHS* in 1999. It was the first scientific paper that clearly showed the massive decline of vultures during the previous 10 years or so. The paper discussed various theories for the cause of the decline, but we needed more research to find out the real causes.

In August 1999, we organized a Vulture Conservation Strategy Planning Meeting, which was attended by 28 scientists and researchers, mainly from BNHS, but also from WII, WWF-India, MoEF, Biodiversity Initiative Trust, Centre of Wildlife and Ornithology, Sanctuary Asia, Bombay Environmental Action Group and BSAP. Based on the discussion, three issues were prioritized for detailed investigation: outbreak of disease, chemical contamination and poisoning. By this time, we were discussing long-term collaboration for bird conservation in general with BLI and the RSPB, the UK-based organization and a BirdLife partner. The RSPB sent a young vet to work with Vibhu but she, like us, could not fathom the problem. Vibhu had seen sick vultures, sitting with their heads down, so we knew something was killing them slowly. Postmortem investigation in Hisar Veterinary College showed that the vultures were suffering from visceral gout. We were told that visceral gout could be due to a virus, so we contacted the National Institute of Virology, Pune, but there was no response from them for a long time.

Later, I had to go there personally and found that they work only on viruses infecting humans.

As the situation was becoming desperate, BNHS organized another meeting in February 2000, for which the RSPB sent Dr Andrew Cunningham, a very respected veterinary pathologist of the Institute of Zoology, London. At the meeting, Dr J.M. Deshpande, director, Entrovirus Research Centre of the Indian Council of Medical Research, Mumbai, suggested that we contact the Poultry and Diagnostic Research Centre (PDRC) of Venkateshwara Hatcheries in Pune. Thus started the collaboration with PDRC, which lasted five years, till the real culprit was discovered in 2004. Andrew and Vibhu visited PDRC and many other labs, conducted postmortems of vultures and field studies, and came to the preliminary conclusion that the *Gyps* species of vultures were dying due to a virus. But this needed to be confirmed by sending the samples to a good virus pathology lab. Andrew suggested sending the samples to Dr Alex D. Hyatt, director of Australian Animal Health Laboratory, for identification of the virus. But the Indian law does not allow sending 'genetic material' outside the country. So, now there was another hurdle to tackle. It took us two years to eventually get permission to send the samples, but unfortunately, by then, the Australian lab had been damaged by a major fire, and remained closed for many months.

By 1999–2000, we had also started writing to the MoEF, GoI, stating that something should be done quickly to arrest the decline of vultures. Thanks to the wide publicity generated in the media by BNHS, by 2000 the vulture problem was known to everyone. However, the MoEF wanted hard scientific data, and rightly so. The comparative population data of Keoladeo National Park from the mid-1980s and the 1990s was not sufficient to convince the MoEF that the problem was widespread, and not just limited to Bharatpur. Fortunately, Vibhu's nationwide surveys conducted in many parts of India between 1990 and 1993, were very useful. The survey was repeated in 2000 in the same areas. The results were shocking. Vibhu and his team found a more than 90 per cent decline in the *Gyps* vultures' population all over the country. Further surveys in 2002 and 2003 showed that between 2000 and 2003 alone, the White-backed vulture had declined by 48 per cent and the Long-billed by 22 per cent. What more could any government want to know in order to act? We also had widespread news and anecdotal accounts of vulture decline, from Gujarat to Assam. Carcass dumps around the country had very few or no resident *Gyps* vultures. Delhi, where in the early 1980s nearly 1,00,000 *Gyps* vultures resided, was left with only a few hundred by 1998–99, and almost none by 2000–01. We had intensive data from Keoladeo National Park from the 1980s, and extensive comparative all India data of numbers from 1990–93 to 2000–03. All of them showed the same result – vultures were dying at a fast rate all over India. But what was killing them?

In September 2000, BNHS organized an international conference, funded by the MoEF and the RSPB. It turned out to be a turning point in the vulture saga, as we all agreed that we had to find out what was killing the vultures, and reverse the decline. The three-day Conference was attended by experts from over 10 countries, including officials from the MoEF, and scientists from the Indian Veterinary Research

Institute (IVRI), WII, WWF-India, SACON, USFWS, BNHS, the RSPB, The Peregrine Fund, BLI, Cambridge University, AMU and many others.

All through 2001–03, we kept gathering more and more evidence that vulture populations were declining all over India, Nepal and Pakistan. The prevailing view at that time was that they were dying due to some infectious disease, most likely a virus, which was spreading from east to west, which is why Pakistani scientists reported the decline from 2000 onwards. The fear was that if the virus spread through the Middle East and from there to Africa, it would also devastate the African vulture populations, the way it was doing in Asia. In this context, in 2000, The Peregrine Fund initiated an Asian Vulture Crisis Project with the Ornithological Society of Pakistan (OSP), and in Nepal with the help of Bird Conservation Nepal (BCN).

Dr Munir Virani, a Kenyan of Indian origin, who worked with The Peregrine Fund, and Martin Gilbert from Cornell University, US, were the team leaders for the fieldwork, and J. Lindsay Oak from the Department of Veterinary Microbiology and Pathology, Washington State University, did the diagnostic investigations. While we were still awaiting permission from the GoI to send the samples to the Australian lab, scientists at The Peregrine Fund took some samples from Pakistan to the US and found no virus or any other pathogen that could have caused vulture mortalities. If virus was not killing them, then what was the culprit? The Peregrine Fund and OSP conducted a survey in over 37 villages in Pakistan. The aim was to find out what was getting into the livestock carcasses. Basically, they were looking for something that was relatively new in the market, something that causes visceral gout in birds, something that was effective and yet inexpensive – and diclofenac sodium perfectly fit that profile. It was a new painkiller recently introduced for veterinary use on livestock in Pakistan (like earlier in India). A quick Internet search by the scientists working on the case indicated that diclofenac is lethal to some species. In May 2003, at the World Conference on Birds of Prey held in Budapest, scientists of The Peregrine Fund presented a paper where they clearly demonstrated that the painkiller diclofenac sodium was causing visceral gout in vultures that consumed livestock carcasses contaminated with the drug.

Shortly after, Bob Riseborough, an American expert, sent me (and others) an email stating the information presented in the paper. As soon as I read it, I literally jumped from the chair and phoned Vibhu, telling him that we probably now knew the culprit behind the mass deaths of vultures. However, it was premature to announce this without more comprehensive data, as it would not satisfy the scientific community, including sceptics, particularly those who had been advocating the virus theory. There were also genuine questions on how diclofenac was reaching vultures to such an extent that it had killed over 90 per cent of the country's vultures in about a decade. The scientists of The Peregrine Fund, Washington State University and OSP did some more experiments, re-analysed the data and published a paper in the prestigious *Nature* journal in January 2004. I would say that this paper was the second turning point in our struggle to save vultures. The beauty of this paper was that it was very simple, with basic statistics (which even I could understand) and clear results. The paper was written by 13 authors, who belonged to six different institutions, with J. Lindsay Oaks as the lead author, and showed 100 per cent correlation between the presence of diclofenac and kidney failure in vultures. After the paper came out, Vibhu and his team also analysed vulture carcasses, which we had meticulously kept in deep freezers. We too found a direct correlation between visceral gout and the presence of diclofenac in vulture bodies. The evidence was all over the place – in research papers, in labs, in widespread dead birds and in diclofenac ampoules in vet shops. Someone had to just knit them together.

This is exactly what Dr Rhys Green, a brilliant field biologist from the Cambridge University, did. In February 2004, while visiting BNHS after attending a workshop at Parwanoo in Himachal Pradesh regarding the Vulture Recovery Action Plan, Rhys asked some fundamental questions: What is the livestock population in India? How many livestock deaths take place every year? What is the lifespan of a vulture (*Gyps* species)? How much food do they eat and how often? How long does diclofenac remain in a live animal, and what happens when an animal that was injected with diclofenac sodium dies? We helped him piece together some of these answers. After knitting all this information together, using some intricate statistical analysis, Rhys came up with a paper in the *Journal of Applied Ecology*,[35] which showed through simulation models that even if less than 1 per cent of the livestock carcasses available to vultures had diclofenac residues (which are lethal to vultures), they would die in the same way they indeed had been dying. Later, through carcass-sampling studies, we found that nearly 10 per cent of cattle carcasses in India had diclofenac residue. No wonder the vultures had disappeared so quickly, leaving our skies empty.

Rhys had come to India to attend an international workshop

in Parwanoo on Vulture Conservation Action Plan. The outcome of this workshop was two major recommendations to save the vultures: to start a vulture conservation breeding programme in India, Nepal and Pakistan; and to immediately ban the veterinary use of diclofenac. Time had come for action, but it was easier said than done. Bombay Natural History Society, in collaboration with other institutes, can give recommendations on the basis of good science, but the implementation of those recommendations is the job of the government agencies.

With the financial support of the RSPB, BNHS appointed a Vulture Advocacy Officer, Dr Nita Shah, to work with varied government agencies, both at the central and state levels. Regular meetings took place with the Drugs Controller General of India (DCGI), MoEF, Ministry of Health, Agriculture Ministry, Animal Husbandry Department, Ministry of Chemicals & Fertilizers, Ministry of Commerce, Veterinary Council of India, veterinary colleges, chemist associations, pharmaceutical companies and Indian farmers' associations to ban the use and manufacture of veterinary diclofenac.

In this saga, the outlook of the pharmaceutical companies turned out to be the most positive. In April 2004, a meeting was organized with the pharmaceutical companies, and we expected resistance from them, as they would lose a key market for their products, but strangely, they said that if the government banned diclofenac they would stop manufacturing it for veterinary use. It took a while for the government to comprehend the gravity of vulture decline and its implications. But when ministries not dealing with forests, species and environmental issues had to be involved, the going got tougher.

Nevertheless, Nita pursued her agenda with a constant series of inter-ministry dialogues, and meetings with top decision-makers. This enabled us to get the green light for appropriate policies and amendments in the Drug and Cosmetics Act. The ban on diclofenac turned out to be one of the fastest decisions taken in a country like India.

The biggest task addressed by Nita in the vulture advocacy programme was to get the policies implemented on the ground. We were battling against time, as the vulture populations continued to crash. Meanwhile, the DCGI came up with the plea that unless there was an alternative to the drug, it would not be possible to enforce the ban on diclofenac. I thought this was strange – it was not as though our livestock had not managed (they had thrived, instead) before this killer drug was released in India in the early 1990s or so, in Nepal in 1996 and in Pakistan in 1998. My main argument then was that when we had banned hundreds of drugs after they were found to have side effects, why couldn't we quickly ban a drug that had already killed more than 90 per cent of the vultures, and which was still killing the remaining vultures at a rate of 40–50 per cent every year. Why were we so selfish and uncaring? Selfish when it comes to human beings, and uncaring when it comes to other species. Are we really the 'thinking animal' (*Homo sapiens*) that we claim to be?

Nevertheless, BNHS, the RSPB, IVRI and other organizations began research on an alternative to diclofenac. Dr Swarup Bhatnagar from the IVRI, Bareilly, came to our rescue and helped in conducting research with his team at his Institute. With all this momentum behind her, Nita was able to get the issue listed for dialogue on the agenda list of the National Board for Wildlife (NBWL) in March 2005.

Only through the intervention of the PM – the chairman of NBWL – did the first circular banning this drug come out in May 2006.

The Asian vulture crisis was so humungous and urgent that the RSPB decided to post Dr Chris Bowden (who incidentally is married to my former AMU student, Dr Farah Ishtiaq) to India, and established a consortium named Saving Asia's Vulture from Extinction or SAVE. In the last 15 years, SAVE has done tremendous work in preventing the total extinction of vultures in Asia and Africa.

The credit for the success of our vulture advocacy campaign must be ascribed to Dr Nita Shah, a multifaceted lady. Besides being a good fieldworker (she worked on the Wild Ass for her PhD), she is a Bharatanatyam dancer, swimmer, skater and a creative artist. She worked with Mike Pandey, the celebrated wildlife filmmaker, to produce a documentary, *The Vanishing Vultures*, which she got translated into eight Indian languages. For many years, we effectively used this documentary for publicity. She also designed posters, pamphlets, stickers and vulture badges. A vulture story, *Jassi Jassus aur Rangeen Atma*, using puppets, was produced in collaboration with Ranjana Pandey – the puppeteer, playwright, theatre and television director. Nita was also instrumental in using the dying art of string puppetry in conservation. She trained a team of Rajasthani puppeteers and built a vulture storyline, which proved to be very popular among urban and rural folks. Nita was thus able to engage various target groups across the country to mobilize quick action in vulture conservation.

Two other names stand out for their yeoman service during the vulture conservation saga – Dr R.D. Jakati and the RSPB – Dr Jakati for his moral and in-kind support, and the RSPB for technical and financial support. Our relationship with Dr Jakati started in 1998, with an innocuous enquiry from the Haryana Forest Department to BNHS regarding a remedy for an injured Cinereous Vulture (*Aegypius monachus*), which the Department had rescued. I had sent Dr Vibhu Prakash, who was still working at Bharatpur at that time, to Haryana. Dr Jakati had heard about the vulture decline, so he suggested Vibhu start a vulture breeding centre. The rest as they say is history.

The result of that suggestion – the Vulture Conservation Breeding Centre at Pinjore in Haryana – is one of the finest vulture breeding centres in the world, thanks to the visionary support of Dr Jakati and his team, and the financial and technical support from the RSPB, Zoological Society of London (ZSL), MoEF and the Darwin Initiative for the Survival of Species, among others. It started as a small vulture rescue centre in 2001, where sick and dying vultures were kept, but now it is a fully developed facility with nearly 450 vultures. Its success can be judged by the fact that now every state wants a similar centre. We then established two smaller centres – one at Buxa in West Bengal, and another at Rani Forest Range in Assam. One more centre came out in Bhopal funded by the MP Forest Department but it was looked after by BNHS.

Despite sceptics' view that BNHS did not have the expertise to breed vultures, in the 2007–08 season, two chicks of Oriental White-backed Vultures were fledged successfully – the first time ever that the species had been bred in captivity. We also benefitted from the regular inputs of a subject expert,

Jemima Parry-Jones from the UK, who has successfully bred over 60 species of raptors. Till now, we have bred nearly 400 vultures in our four centres, and nearly 100 have been released. More will be released in 2024.

The vulture conservation programme also demonstrates the importance of collaboration and capacity building. BNHS's philosophy is to work with the government while retaining its independence. Our collaboration with the governments of Haryana, West Bengal, Assam and MP, and also the MoEF (renamed MoEFCC since May 2014), proves that if NGOs and government come together, we can achieve a great level of success. In its long history of 140 years, particularly after Independence, BNHS has worked with several international agencies, such as IUCN, BLI, USFWS, Smithsonian Institution, Yale University, World Pheasant Association, WWF, Wetlands International, International Crane Foundation and many more.

A key partner in the vulture advocacy programme was the RSPB, whose support came to us, as they were a major BirdLife partner. The vulture conservation breeding programme in India and Nepal is one of the biggest overseas programmes of the RSPB. Besides funding the three breeding centres and supporting the advocacy programme, the RSPB's significant contribution comes from their technical support, in the form of experts like Chris Bowden, Debbie Pain, Rhys Green and Richard Cuthbert; they also helped us connect with other experts, such as Andrew Cunningham, Andrew Routh and Nick Lindsay of ZSL, and Jemima, founder of International Centre of Birds of Prey. Jemima is one of the most famous experts on captive breeding of raptors, so her help in designing the cages, and sharing with us the processes of veterinary care, feeding, chick-rearing and animal welfare was tremendous. She is now the chairperson of SAVE, and is always available for any assistance required by our conservation breeding centres.

The support of all these agencies and people has made the vulture programme the longest-running initiative of BNHS in its 140-year history. However, the success of this multi-institution programme can be judged only when our skies are once again filled with these majestic masters of the sky.

19

Field Visits

In the last 50 years, I have travelled across almost the whole length and breadth of India. Some states, such as Maharashtra, MP, Rajasthan, Gujarat, Assam, Jammu & Kashmir and Sikkim were visited many times, and for longer periods, due to the nature of my work. While field visits were many, and descriptions of all of them will probably fill a book independently, three in particular stand out.

India has two types of major islands – the Andaman & Nicobar Islands located in the Bay of Bengal are mainly peaks of a submerged mountain chain, comprising more than 572 islands and islets, with a total geographic area of about 8,200 sq. km; while the Lakshadweep archipelago consists of 36 coral islands covering 12 atolls, with a total area of 32 sq. km. Only 11 islands are inhabited, of which Agatti is the most populated. These islands are irregularly scattered in the south Arabian Sea, about 280–480 km west of Kochi, off the Kerala coast.

I had had the opportunity to visit almost all the important islands of Andaman & Nicobar, but I had not seen Lakshadweep. That opportunity came in 2007, when I had to attend a workshop in Kerala. After the workshop, from Cochin, I took a flight (a small 16-seater Dornier) to Agatti. As the helicopter flight to Kavaratti Island, my final destination, was four hours later, I spent a few hours on this wonderful paradise, where I was also witness to the only recorded accident in the last few decades. My diary entry for that day reads, 'Few autos and vehicles. On the way saw an accident – a minibus rammed into a coconut tree, bringing down some coconuts, one injuring a passerby. I thought I had left all such accidents behind in Bombay, but I was wrong.'

In the 2000s, BNHS had a major project on the Giant Clam (*Tridicna gigas*) and other aspects of the marine biodiversity of Lakshadweep, funded by the Darwin Initiative for the Survival of Species. Deepak Apte was the principal investigator of the project, which he had been instrumental in securing for BNHS. The project was executed almost entirely by the locals; Deepak had six young budding scientists in his team, who not only counted the number of Giant Clams in the lagoons, but also cooked excellent tuna fish. He had developed a wonderful rapport with the local people, from the traditional boat-maker to the lighthouse man. As we were working side by side with the local people, building their skills, training them and also learning from them, BNHS became popular in Lakshadweep. Wherever I went, and in whomever I met, I only saw admiration for BNHS.

Thanks to Deepak, I spent seven memorable days in Lakshadweep. But I kept thinking of an impending disaster – if the climate change goes on at the present rate, the oceans will rise and there will not be any Lakshadweep left, as the islands are just 1–1.5 m above sea level. This was the only inconvenient truth that spoilt my 'holiday', as I sat in the serene palm-fringed beach, looking at the blue horizon.

One day before I returned to the mainland, I visited Pitti Island, an uninhabited sandbank, about two hours by boat from Kavaratti to the middle of nowhere (the key reason for its survival!), where thousands of birds, mainly Sooty Tern (*Sterna fuscata*), Great Crested Tern (*S. bergii*), Bridled Tern (*S. anaethetus*) and Brown Noddy (*Anous stolidus*) breed. I had never been to a seabird nesting colony, so my visit to Lakshadweep would have been incomplete without going to Pitti Island. An excerpt from my diary of 27 January 2007 reads:

> Two-hour journey to the Pitti Island was uneventful, and somewhat boring as all around was sea and nothing [else]. At a distance, some terns [were] foraging, indicating [the presence of] tuna [...] From [a] distance, Pitti was [like] a line of white ribbon on the surface of [the] vast blue sea. As we approached closer, we encountered the famous strong waves, the reason why Pitti is so inaccessible. From the larger boat, we shifted to a smaller one, [which] could [get us] closer to the island. As we neared the island, Idress, Hussain, Raheem, Deepak and others jumped from the boat into the sea. I also jumped out into the low surf, and they all pulled me to the shore, before a big wave could reach us. It was fun jumping from the small boat into the surging sea and then swimming/running towards the shore, with everyone trying to help me.

Once we were on the island, we were safe from the hard-hitting waves. The island was replete with birds, mainly terns and noddies. After spending two hours, we left the island, reversing the process of arrival – we ran to the small boat before large waves could hit us, then transferred from the small boat to the larger boat, climbing through ropes in a choppy sea. One of the party slipped into the sea but fortunately he knew swimming, and managed to scramble onto the smaller boat before clambering aboard the larger vessel. The art of swimming being unknown, I reassured myself that had I had fallen into that choppy sea, there were enough *jawans* around to rescue me.

It was a memorable trip and one that I still vividly remember, even 17 years later. The highlight was the sighting of Pomarine Jaeger (*Stercorarius pomarinus*), a large predatory bird with a 56 cm wingspan, that is a nemesis for eggs and chicks of terns. Despite many plans, somehow I could not visit Lakshadweep again – a regret that I wish to address one day. As the age of 74, I still have the spirit to take risks, such as jumping into a choppy sea to revisit my beloved birds. The *jawans* are now 17 years older, but they too have the same enthusiasm and energy. I am in touch with a few of them, and if an opportunity arises, they have promised to take me to Pitti Island once more.

In 2008, I was in the extremely beautiful Overa-Aru Wildlife Sanctuary in Kashmir, where one of my former students, Syed Suhel Intesar, was in charge as the wildlife warden. I had visited the Aru area three years ago and fallen in love with the place. It is one of the prettiest places I have ever seen in my life – a place of classical picture postcard beauty, much like we see in the photographs of Switzerland. I first heard about the Overa-Aru Sanctuary through the papers published by Trevor Price, an American ornithologist, and our very own Nitin Jamdar, an active member of BNHS, and now a judge of the Bombay High Court. For five summers, they studied the breeding biology of the warbler species in this Sanctuary. One of their papers was published in the *JBNHS*. Overa-Aru has since been identified as an IBA.

With three of my former students from the Aligarh days, Syed Suhel Intesar, Maqbool Baba (both with the Wildlife Department of the Government of Jammu & Kashmir) and Khursheed Ahmad (who works in the Faculty of Veterinary Sciences and Animal Husbandry, SKUAST), I visited the Aru part of the Sanctuary in June that year. The route to Aru runs through Pahalgam, a famous tourist place. Pahalgam is a classical case of how *not* to develop a place! For a nature lover like me, Pahalgam is all but destroyed, with crass and hedonistic mass tourism – cheap hotels, restaurants blaring out loud music, streets overburdened by wrongly parked vehicles, ugly hoardings, flashy nouveau riche tourists and non-degradable litter everywhere. Everything that could go wrong has gone wrong in Pahalgam, much like in other hill stations, such as Ooty, Mussoorie and Nainital. Even a dreadful water park has been developed close to the beautiful Lidder River, on the insistence of a former CM. We Indians have developed an uncanny ability to destroy the places that we want to 'develop' to attract tourists. I would not recommend that anyone with a tinge of interest in nature should visit Pahalgam. However, 12 km away, the Aru valley is just the opposite, having not yet suffered the onslaught of mass tourism.

Many conservation activists criticize the Forest Department for various omissions, but have they ever thought what would be the status of wildlife and 'wild' places without the Forest Department? Fortunately, Overa-Aru is under the Wildlife Department. In Jammu & Kashmir, there is a full-fledged Wildlife Department, separate from the Forest Department, unlike in other states where the wildlife department is within the forest department. Many conservationists of India, particularly our indomitable Valmik Thapar, have been insisting that protected areas should be managed by a separate wildlife department. Countries such as the US have the National Park Service and the Fish and Wildlife Service, which look after the wildlife areas, while the US Forest Service manages national forests and grasslands, which are equally important for wildlife conservation.

Although Aru area is only 12 km from Pahalgam, not many tourists visit it. Moreover, the Wildlife Department has constructed a gate at the Sanctuary's entrance, and tourists have to pay a fee to enter – Rs 25 for Indians and Rs 50 for foreigners (local inhabitants of Aru are exempted from the entry fee). This controls the traffic to this pristine Sanctuary.

Although there are many private guest houses, mainly homestays, with basic bed and breakfast facilities, hotels are missing (fortunately), except a beautifully built government tourist guest house. Many Bollywood movies have been shot here, including the popular Dilip Kumar starrer *Karma*. The locals still excitedly remember the visit of the famous thespian, although it happened a long time ago. They will tell you where Dilip Kumar sat, what he did, where he looked, etc.

Aru is a small village of about 45 houses, situated on the banks of the crystal-clear Lidder River, which originates from the Kolahoi Glacier. For almost six months of the year, it is covered in snow and is mostly inaccessible to visitors. But summer brings life, in the form of millions of flowers, insects and birds. It also brings some tourists, who mostly take a pony ride to visit the beautiful vales of Koot Pathri, Katar Nag and a high-altitude lake Tarsar. Before the insurgency in Kashmir, Aru was a very popular destination for trekkers. Thankfully, peace is returning to the Valley, bringing back the tourists and visitors, and ironically some trouble (in the form of non-degradable litter).

The people of Aru are financially poor, but extremely hospitable, friendly and good-looking. Due to the difficult and hilly terrain, there is not much cropping in the Aru village, so villagers eagerly look forward to receiving tourists; there are nearly 100 ponies available for hire by visitors. Until a few years ago, pony owners would fight among themselves to offer their services to tourists, which sometimes even resulted in physical assaults and also undercharging. However, soon the villagers realized that by competing among themselves for the limited number of tourists, they were losing out. With the wise advice of the village elders, the pony owners decided to form a sort of cooperative. Now, each pony is registered and given a number. If 50 ponies are used by tourists on a given day, the next day the remaining 50 take tourists around. But the most interesting part is that the earnings of a day are pooled and divided equally among 100 pony owners. So, even if an owner's pony is not used, he would still get around Rs 200–300 a day (depending on the tourist influx). Pony charges are fixed (though they vary for different routes), and no one is allowed to undercharge.

Along with Intesar Suhel, Khursheed Ahmad and Maqbool Baba, I met Aru's Sarpanch, Gulam Mohemmad Malik, and discussed the polythene pollution problem that had reached even this remote, once pristine village. He was also very concerned about the village becoming dirtier day by day, but expressed his helplessness. Citing the example of the successful pony cooperative, I requested him to do some *shramdaan* with the help of the villagers and collect all the plastic littered around. The Sarpanch agreed to my suggestion and added that perhaps some NGOs too could help in cleaning this otherwise extremely beautiful village.

The biggest problem that the villagers of Aru face stems from the Wildlife Protection Act. Aru village is a part of the Sanctuary, and so villagers are not allowed to freely repair or extend their dwellings, or expand their small crop fields. Actually, this is a blessing in disguise as far as tourism is concerned. I also told them that if Aru is allowed to 'develop' like Pahalgam, their lifestyle and values would be destroyed. 'Development' of the area would not necessarily mean that the local youths would get good jobs, as commercial tour operators and hoteliers will employ professionals from the hospitality industry. Most of the villagers understood this. They told me that some rich hoteliers and prospectors had already bought land around Aru, in anticipation of convincing the government to denotify Aru from the Sanctuary. Most of the villagers with whom I spoke, including the Sarpanch, agreed that only the fact of their village being a part of the Sanctuary had kept the land sharks at bay. However, they want some relaxation in the Wildlife Act, so that they are allowed to repair their buildings or extend them within their own lands. This, I think, is a reasonable demand, one I hope the state government accedes to.

In 2013, Dhritiman Mukherjee, one of India's ace nature photographers, and I decided to go to Andaman's Narcondam Island to study and photograph the Narcondam Hornbill (*Rhyticeros narcondami*). As Narcondam Island is primarily a wildlife sanctuary, the permission to visit the island is given by the Forest Department, which was quite easily obtained. I told the forest officials that my assistant would accompany me, not revealing Dhritiman's name, mistakenly thinking that the Department as well as Shirish Manchi (from SACON, and a former PhD student of the late Dr Ravi Sankaran), who was organizing the visit, would not like a photographer tagging along with me. The difficult part of the trip was reaching the island – about 142 km from Diglipur and 189 km from Port Blair, the capital of Andaman & Nicobar. The small, isolated, 6.8 sq. km volcanic island (the volcano is dormant though) can be reached only by hiring an expensive yacht (which was certainly out of the question for me) or through the Coast Guard that patrols our territorial maritime zone. Since Narcondam Island is quite close to Myanmar and Thailand, the movement of illegal fishers into India's territorial waters is quite common. In the mid-1960s, after the GoI got Narcondam Island in exchange for Cocos Islands from Myanmar, it was decided to post policemen on Narcondam to keep poachers away. These policemen stay on the island in one-month shifts; the Police Department does not have its own vessel, and so the Coast Guard helps them manage the movement of personnel. Most of the researchers who have been to Narcondam have gone there with the help of the Coast Guard. I was working at my office in Hornbill House one afternoon, when I got the news that the Coast Guard had agreed to take me, along with Shirish and my 'assistant'. I jumped at this once-in-a-lifetime opportunity. All official commitments for the next 15 days were postponed or cancelled.

One phone call to Dhritiman was all it took – he too cancelled all his commitments, including telling Zeiss, the German manufacturer of optical systems and optoelectronics, that he would not be able attend the event to receive the Carl Zeiss Conservation Award on 23 March, as photographing the Narcondam Hornbill was far more important to him than the award! We were informed that the Coast Guard ship would sail from Port Blair on 17 March. Shirish was the first to reach Andaman, as he still had a lot of paperwork to do for our trip; Dhritiman reached on 14 March and I on 16 March. On 17 March 2013, the three of us sailed along with 17 policemen at about 11 p.m. on the Coast Guard ship. Dhritiman and Shirish hit it off quickly, as if they had known each other for donkey's years. A perfect Luv and Kush!

The ship was commanded by Captain Manjeet Singh Gill, whose wife is a BNHS member. Both of them are interested in wildlife, and so the next morning we had a lovely time talking to Manjeet Singh about various conservation issues, including those relating to Narcondam Island. By 8 a.m., we could see the island. The Coast Guard ship being large was anchored about one nautical mile from the island, and we were transferred to the island in slow-moving dinghies, affectionately called *jimini*. By noon, the new police team was in place, while the old team (after completing their month-long shift) was all set to be taken back to Port Blair on the Coast Guard ship. Like returning school children, the old team was quite happy at the prospect of a reunion with families and friends, while the new team was slightly depressed at the thought of spending a month with no contact whatsoever with their near and dear ones. One young man, who had got married recently (in February), was particularly depressed at parting from his young wife so soon – and so he was the focus of many efforts to boost morale! By evening, the new group of policemen were ensconced in their primitive living quarters. They gave me a cot in the dispensary, which they felt would be more suited to my 'stature' while Dhritiman and Shirish shared a small derelict room, not used by the policemen. The compounder was a genteel man and he tolerated my presence with a composure that comes with experience.

For the next 14 days, we not only studied the Narcondam Hornbill but also saw the gentle side of the police corps. They have a lovely custom at Narcondam – the outgoing police party prepares food for the incoming group; another custom is to provide sweet drinks to anyone who arrives on the island. For instance, when a boatload of 24 labourers came to repair a broken pipeline and some parts of the decrepit camp quarters, the policemen greeted them with *sharbat*.

Another thing that I noticed was the bonhomie and friendliness of the team. In the 14 days that I was there, I never heard anyone talking loudly or using swear words (not uncommon with policemen on the mainland); neither were there any fights. All of them were rather young and the average age must have been 25 years; their untamed energy was spent in doing mundane household chores, such as fetching water, cleaning the camp, washing clothes, picking out the small pieces of stone from rice and dal, and playing volleyball in the evenings. As electricity was provided for two hours (post sunset), that precious time was used to watch dubbed Hollywood movies, while we used the opportunity to recharge the camera and laptop batteries, and to download the day's pictures.

Besides a compounder, a cook (who had earlier worked at the famous Megapode Hotel in Port Blair) lived in those quarters. Every day he proved his culinary skills with the limited food items that they had managed to bring from Port Blair. Rice–dal and dal–rice were the 'two dishes' on the menu, with an occasional fish or crab! The ingredients that he lacked were made up for by the affection with which he fed the hungry policemen, and us as well. Lunch and dinner were the occasions when we all sat together and literally ate from the same plate – which created a strong bond between us.

Twice I was woken up softly at midnight to celebrate the birthday of a policeman. When we had all taken up our 'positions' silently, armed with a cake made of *atta*, on which candles had been hastily placed, the birthday 'boy' was woken up to confront half-naked and half-sleepy colleagues singing 'Happy Birthday'. Such small incidents will remain etched in my memory forever.

Dhritiman spent the days working hard at the craft of photography; the more I saw his images, the greater was my admiration for this young man. He is a class apart. I consider him not a wildlife photographer but an artist who uses a camera creatively. Dhritiman is a rock climber, trekker, diver and a very committed conservationist. Similar to a field scientist, he can sit for hours in the hide, quietly observing the behaviour of his subject before clicking a picture. In the same spirit of focus, he will even hang from a tree to get a particular angle. Unlike many photographers, Dhritiman never compromises on the welfare of animals; on a particular Narcondam Hornbill's nest, an irritating leaf would hide the face of the male when he came to feed the female and the chicks. We had the choice of removing the small branch to get clear shots, but Dhritiman refused to do any 'management' of the nest. Fortunately, after a few days, the leaf fell off naturally, giving us an opportunity to take unhindered pictures. All the pictures were taken when the birds were totally relaxed, and going on with their normal activities. It was a pleasure to work with Dhritiman in Narcondam. Besides studying the hornbills, I was also studying Dhritiman, bringing to mind the famous book *Man Watching* by Desmond Morris.

While not photographing, Dhritiman would spend his time going over the pictures, deleting what he felt were not up to the mark and improving the remaining ones, all the while singing Urdu ghazals and songs in his heavy Bengali accent. I would repeatedly correct his Urdu pronunciation, but he would unlearn the correct ones as quickly as I taught them! In those 14 days, I could not teach him Urdu, but I picked up some Bengali words, another beautiful language of our country.

In this chapter, I have described visits to three geographically and ecologically distinct areas of our country. Although geographically diverse, what connects them is their remoteness and inaccessibility, which seems necessary to save wildlife from 'development'. As my idol, M. Krishnan, who was also an ecological patriot, wrote in his last column printed on 18th February 1996, '…the identity of a country is dependent not so much on its mutable human culture as on its geomorphology, flora and fauna, its *natural* basis.' Working in BNHS gave me opportunities to visit the length and breadth of our beautiful country, see amazing wildlife and meet extraordinary people. I hope to describe my fieldwork in greater detail in another book.

20

More Feathered Friends

Research and surveys under the GIB and florican projects gave me ample opportunities to visit almost the whole of India (except Kerala, Kashmir, Punjab, Haryana and Odisha, where these birds are not found) and collect data on many associated species. For example, while conducting surveys of the GIB and Lesser Florican, I collected data on Chinkara and Blackbuck – the two species that share the habitat with bustards – which led to the publication of three scientific papers[36–38]. Although there were some papers on Blackbuck, there were no good papers on Chinkara, so my three papers became a sort of baseline information on this species. I also collected data on the Indian Grey Wolf, which was not enough to write a paper or report, but I shared my unpublished data with Dr Y. Jhala and my first PhD student, Dr Satish Kumar Sharma, who worked on wolves. As my first love is birds, I also collected data on all birds during my surveys of bustards and floricans.

During the surveys, if I came to know that there was a wetland close by, I would visit it and collect data on all major bird species, such as storks, cranes, eagles, geese and ducks. During my frequent visits to Keoladeo National Park, I realized that despite the fact that the *jheel*s extend to nearly 9 sq. km (when full), there were only five to six pairs of Black-necked Storks, and all the pairs were found discretely. Firstly, I found that they are highly territorial even in the non-breeding season, and secondly, that they occur in low density in good-quality wetlands, which have plenty of large fish. I collated six to seven years data of stork sightings from all over India and wrote a paper on the status of the species in India, which was published in *Forktail*,[39] a British journal in 1989.

While surveying UP, Bihar, West Bengal and Assam for the Bengal Florican, I was fascinated by the two adjutant storks found in India. The result was a joint paper with Goutam Narayan and Lima Rosalind, in 1990, on the Greater Adjutant (*Leptoptilos dubius*).[40] Based on these two papers (on the Black-necked Stork and the Greater Adjutant) and some more work, I led a research project on Indian storks that was funded by the USFWS, which became the vehicle for doctoral studies for three of my students. But doctoral studies and writing papers were not my main aim, highlighting the plight of wetlands and fish-dependent species was. I am happy that subsequently many researchers have worked independently on these stork species, particularly Dr K.S. Gopi Sundar. Based on these studies, we now realize that the presence of large storks and their numbers and diversity are a good indicator of the habitat quality of a wetland. After reading my papers and interacting with me, Arvind Mishra of Bhagalpur, Bihar, conducted exemplary community work on the protection of Greater Adjutant, resulting in a three-fold increase in the population of these endangered birds. In Assam, Dr Purnima

Barman's work involving villagers is well known globally; it also won her the 'Green Oscar' in 2024. I cannot take credit for her achievements but a spark was lit by me 30 years ago, when I had highlighted the plight of these majestic birds.

The Swamp Francolin (*Francolinus gularis*) was considered a rare species, so while working on the Bengal Florican, we collected data on this species as well, and found that although uncommon it was not as rare as thought earlier. It is a species that is totally dependent on tall, wet grasslands of the North Indian terai and the Brahmaputra floodplains, but fortunately, it has adapted to live in the sugarcane fields of the terai. During my surveys in UP, West Bengal, Assam and Arunachal, I would collect data on the Swamp Francolin and subsequently, Salim Javed, Qamar Qureshi (of WII) and I published a paper on the species.[41] The World Pheasant Association was interested in this species, and while working in AMU we did a small project on Swamp Francolins, with funding from their 'Quail, Partridges and Francolin' Working Group. In 1994–95, at the Dudhwa National Park we radio-tagged a few Swamp Francolins and studied their movement and nesting success. This three to four month study resulted in a paper published[42] many years later.

Another species that got written about as a result of those surveys was the White-browed Bushchat (*Saxicola macrorhynchus*); my two papers were the first ever on this species.[22, 23] I am happy that now it is on the bucket list of birdwatchers visiting the Thar Desert. The White-browed Bushchat, seen mostly in the arid habitats of north-west

India, gets its name from its very noticeable white eyebrows. What attracted my attention to this relatively unknown bird in the early 1990s was that practically nothing was known about this species at that time. Consulting the magnum opus of Indian ornithology, the 10-volume *Handbook of the Birds of India and Pakistan* by Dr Sálim Ali and Dr Dillon Ripley, revealed that it is found in the Thar Desert and the nearby drier areas of Punjab, Haryana and western UP. The book specifically mentions Aligarh, where I was teaching back then (from 1991 to 1997). Aligarh has some drier areas, so we went looking out for this bird, but found none. What further tickled my curiosity was that although I had conducted many surveys in the Thar Desert in the 1980s, I had never come across the White-browed Bushchat. In 1992, my friend, Dr T.J. Roberts mentioned in his book *Birds of Pakistan* that Stoliczka's Bushchat was extinct in Pakistan, as he could not find any evidence of the species during 28 years of fieldwork in that country. Quick correspondence with the BLI and Oriental Bird Club, UK, brought the good news that Dutch ornithologists had photographed it near Khara, a village between Phalodi and Jaisalmer. I had gone through this village many times during my surveys of the GIB but I had never noticed it, perhaps due to my preoccupation with the bustard.

With the support of Oriental Bird Club, Cygnus Wildlife Holidays and WWF-India, I decided to uncover the mystery of this 'extinct' bird. The first survey, spanning over three weeks in 1993, revealed four birds between Undu and Kanasar in Jaisalmer and Barmer districts, respectively. None were seen in Gujarat, not even in Rapar, where the first specimen was collected 120 years ago. In the second survey that year, only one could be seen in four weeks. The third survey in January–

February 1994 delivered the bonanza, and we saw about 81 individuals across fifteen 15 sites in Rajasthan. Once my papers and articles were published, many more people got interested in the bird, and now its presence has been recorded in Sultanpur Bird Sanctuary and Hisar in Haryana; Naliya in Kutch; Velavadar in Gujarat; Little Rann of Kutch; Desert National Park in Jaisalmer and Barmer; Diyatra in Bikaner; Keoladeo National Park in Bharatpur; Khichan in Jodhpur; Ranthambore in Sawai Madhopur; Sonkhaliya in Ajmer and Tal Chhapar in Churu. It is also reported from Akola near Pune, Maharashtra. When I started work on this species, there was not even a good picture, but now we have a surfeit of images, thanks to the long-lensers on the lookout for rarities. The portal *eBird* has 1039 records, with many repeated from the same sites by different birdwatchers.

While studying this bird, I noticed an interesting aspect in its behaviour – it does a puff-and-roll dance while foraging on the ground, which does not appear to be related to aggression or territoriality. While looking for insects, the bird puffs up the breast feathers and sways sideways. The whitish breast and white belly become conspicuous and the bird appears larger than normal. I always recorded such behaviour while the bird was foraging on the ground, but Nikhil Devasar, a very competent birdwatcher, saw a female puffing from a high perch when he first approached it for a photograph. Like most members of *Saxicola*, the White-browed Bushchat perches on a small bush and looks all around. When a prey is sighted, it descends and picks it up, a swoop sometimes, preceded by the puffing dance. We still do not know the purpose of this puff-and-roll display. Is it to startle the insect? But why would the bird do this when it has already seen a prey from its perch? It is an interesting topic for young researchers to study. Another curious behaviour I noticed was that some individuals were quite bold and allowed you to approach close (5–10 m) to them, while the others flew away even if the observer was 50–100 m away.

Despite all this research, the White-browed Bushchat is still a mystery bird to me. We do not know where it nests, how many eggs it lays, where does it go in certain seasons and what its ecological requirements are. Its cousin, the White-throated Bushchat (*Saxicola insignis*), breeds in a small area in Mongolia and Russia, and migrates in small numbers to North India, Nepal Terai, Northeast India and Myanmar. The White-browed Bushchat is not seen throughout the year, but only mainly during winter months. I wonder whether it too migrates to India from some unknown areas of Russia or Mongolia? Fortunately, we now have geolocators and GSM-based systems for tracking small birds. I hope future researchers show an inclination to study this enigmatic Indian bird.

As the cliché goes, the Western Ghats is a biodiversity hotspot of India. Though the habitat is biodiverse, it's under great threat of destruction. As far as birds are concerned, 26 species are endemic to the Western Ghats. The Nilgiri Laughingthrush (*Montecincla cachinnans*, earlier termed *Strophocincla cachinnans*) is a Western Ghat endemic that is confined to a limited area of around 300 sq. km.

In India, there are 30 species of laughingthrushes, mostly confined to the Northeast and the Western Ghats. The Nilgiri Laughingthrush has perhaps the smallest distribution among its relatives. As the name indicates, it is only found in the higher reaches of the Nilgiris and the adjoining hill ranges. Due to its limited distribution range and ongoing threats, IUCN has listed it as 'Endangered' in the Red List. The bird is confined to shola forests (tropical montane forests comprising grasslands and stunted trees), which are a characteristic feature of the southern Western Ghats. During the last 200 years, the shola forests have seen various threats, such as clear felling to convert them into tea gardens or plantations. Invasive plant species and overgrazing are other major threats.

Based on the habitat suitability assessment, Ashfaq Ahmad Zarri, my PhD student, found that the total area occupied by this endangered bird is no more than 268 sq. km. This area comprises 584 highly fragmented patches (natural as well as man-made), distributed all over the Nilgiris. The smallest suitable patch identified by the habitat suitability model was 31.8 sq. m and the largest was 7.5 sq. km. Also, 80 per cent of these patches were smaller than 0.5 sq. km. Our study showed that the Nilgiri Laughingthrush has a much narrow area of occupancy than thought earlier. The total world population of this restricted range species could be less than 2,000 individuals.

Mukurti National Park (78.46 sq. km) is the only protected area in the Upper Nilgiris. In the rest of the Upper Nilgiris, the natural habitat of the Nilgiri Laughingthrush is practically unprotected, and thus faces various threats. Based on the above study, we recommended that the size of the Mukurti National Park be increased to include Avalanche, Upper Bhavani and Kundah forest ranges located on its east. But this proposal is long pending clearance from the government. We also identified many other shola forests, such as Naduvattam,

Bison Swamp, Governor's Shola, Cairn Hill Reserve Forest, Longwood Shola and Thiashola, as important habitats of this endangered bird. Organic farming (which is as important as habitat protection) that does not use harmful chemical pesticides should be encouraged and incentivized in the Nilgiris Laughingthrush's habitat and surrounding areas; if necessary, organic farming should be subsidized and crop insurance of inorganic farms should be discontinued. The Nilgiri Laughingthrush uses wattle plantations adjoining the sholas as corridors for movements between the shola patches. If it is not possible to restore the sholas in areas where they have been cleared for plantations, the plantations should be preserved so as to minimize the fragmentation of populations of this endangered bird. Fortunately, the shola forests are protected by law, but smaller patches of less than 500 sq. m are under intensive pressures from the surrounding human-dominated habitats. Recent increase in tourism in the Nilgiris has added biotic pressures, and some small patches of the shola forests may not survive, but thankfully, Mukurti National Park is still safe from further fragmentation.

From the sholas, we move to the distant shores of the Andamans. My association with the Edible-nest Swiftlet (*Aerodramus fuciphagus*) started with Ravi Sankaran, and is still continuing with his student, Shirish Manchi, after the unfortunate death of Ravi in 2009. During his lifetime, I could not visit the field stations where Ravi worked on these small swiftlets that nest in dark caves, but I got the opportunity to visit the Chalis Ek Caves in 2012, where Shirish's student, a young lady named Akshaya Mane, was working for her PhD. A visit from her mentor's mentor excited her. Despite my claustrophobia and fear of narrow dark spaces, Akshaya managed to convince me to enter the narrow cave and see the nests.

Many swiftlet species produce edible nests that are used in East Asia to make soup. The Edible-nest Swiftlet, in particular, creates a small cup-shaped nest from its saliva. Such nests are in high demand in Indonesia, Malaysia, China and Hong Kong, where they sell for US$400 per kilo. In India, there's another variant of swiftlet – the Indian Swiftlet (*Aerodramus unicolor*), which is found in the Western Ghats and Sri Lanka. It builds a similar nest, but adds small twigs and feathers to it, rendering the nest inedible and therefore not so valuable.

The Edible-nest Swiftlet is endemic to the Andaman & Nicobar Islands. Ravi had conducted extensive surveys all over the Islands and concluded that conventional methods cannot protect the nest caves from poachers, who were more knowledgeable and innovative than the forest guards, and more determined too. In those remote caves, poachers have to be successful only once in order to harvest all the nests, while the guards have to be vigilant all the time. And so, the odds were in favour of the nest poachers. Also, the locals had no interest in protecting the bird, as they had no benefit from the nests. Ravi had suggested that the government allow sustainable harvesting of nests by locals or a cooperative agency, after the breeding was over. That way, neither adult birds nor young fledglings would be harmed. Since swiftlets build a new nest every year, there is no harm in harvesting the used nest. Since the swiftlets were listed under Schedule-I of the Indian Wildlife Protection Act in 2002, no part, including used nests (after the chicks have flown) could be sold. It took us 10 years and many meetings to 'temporarily' delist it from Schedule-I, so that a new type of conservation paradigm could be applied for this species. According to the new policy, implemented in 2012, villagers were allowed to sustainably harvest nests.

In February 2019, I returned to the Andamans on the invitation of Shirish Manchi (now a faculty member of SACON), who is continuing the work of Ravi Sankaran. At the Islands, he was involved in the counting of Edible-nest Swiftlets that were roosting in a deep cave on Interview Island. I had not been to the Interview Island earlier, so I jumped at the offer. In February 2019, we set off early one morning from Mayabunder to the Chengappa Bridge over the Austen Strait, which separates the Middle and North Andaman Islands. At the Strait, we shifted to a small motorboat to reach Austen Creek, the largest creek in the Andamans. It took us one and a half hours to get to the Creek on that small boat. On both sides of the Creek were fabulous mangroves, followed immediately by tall, dense forests. We reached Interview Island in about three hours, and after exchanging pleasantries with the policemen and forest staff, we walked 13 km to reach the swiftlet protection camp. While I was engrossed in making notes of the endemic birds that I could hear and see, Shirish (and his student who accompanied us) cautioned me about the 9–10 'wild' elephants that occupied the 134 sq. km island. Although we saw dung piles in many places, we did not encounter any elephants. When colonial timber operations were underway, elephants were brought in to move the lumber. Another animal introduced by the British in the Andamans (including on Interview Island) was the Chital or Spotted Deer (*Axis axis*). We heard a few Chital calls and saw pellets in many places but did not see the animal. Interview Island is now a wildlife sanctuary, with tree cutting banned four decades ago. The small elephant population roams free and sometimes troubles the unwary forest and coast guard staff.

Hot dal, rice and sabzi greeted us when we reached the field station past noon, totally exhausted from the long trek in the hot, humid forest. This was followed by a quick post-lunch nap. In a tropical, forested island, with temperatures of 35–40°C and 80–90 per cent relative humidity, an early afternoon siesta is a daily routine. But it had to be a short one, as in that far eastern corner of the country, the sun goes down by 5 p.m.

Shirish and the Forest Department had numbered the caves. Shirish explained that White-bellied or Glossy Swiftlets (*Collacalia esculenta*) and Edible-nest Swiftlets roost in the same cave, but the Glossy Swiftlets return before dusk, as they cannot echolocate (navigate using reflected sound). We sat in the forest in the fading light, slightly away from one of the entrances to the swiftlet cave. Shirish's students, whom we met at the camp, and the other swiftlet protectors went to collect data from the cave's second mouth. As dusk fell, Shirish began counting the number of Glossy Swiftlets entering the cave; his task was made more difficult by some individuals flying out and returning. The counting was just a prelude, for soon it was time for the grand finale – the purpose for which I had come to the Andaman Islands again. In total darkness, when I could not even see my own hands, the first Edible-nest Swiftlet arrived. We started counting the 'tick-tick' sound that indicates the arrival of an Edible-nest Swiftlet, and even the loud call of the endemic Hume's Hawk-owl (*Ninox obscura*) could not divert our attention. First, they came in a trickle, but after about 20 minutes, the frequency increased;

then within 30–40 minutes, almost 70 per cent of the cave's population had entered the roost site. Sometimes three to four came simultaneously, so only an experienced person like Shirish could count them. Till 7.50 p.m., we counted about 175 Edible-nest Swiftlets entering the cave. After an hour, when the 'tick-tick' stopped, we walked back to camp, about 50 m away, with Shirish's headlamp guiding us.

In many countries the Edible-nest Swiftlets are extensively farmed; their breeding colonies are protected and nests are harvested after the breeding is over. In Indonesia, Malaysia, Thailand, Vietnam and more recently, in Cambodia, special 'houses' have been built and modified to simulate the conditions of a natural cave to encourage swiftlets to nest in artificial nesting colonies. Research has shown that we can create conducive conditions that simulate natural limestone caves. Many large nesting colonies have also been established in modified houses where villagers protect the birds and harvest nests at the beginning of the season, so the birds make another nest to raise their chicks. Once the chicks have flown, the used nests are also harvested. Like many colonial nesting birds, Edible-nest Swiftlets come back to nest in the same area again and again, making a new nest every breeding season.

Many countries have made such sustainable use of nests and turned it into a cottage industry. As a result, there are millions of free-flying swiftlets. In India, however, the debate is still on about whether to allow sustainable harvesting or not. While villagers and locals have no interest in protecting the nests, dedicated forest guards look for direction from their bosses; research work too suffers from a lack of funds. The best way to protect the nests from poachers is to commercialize its sale, so that villagers benefit from it economically and continue supporting the swiftlet conservation plan. Ravi Sankaran's excellent plan, later strengthened by Shirish's research, and my strong letter of support to the MoEF, all remain buried in the government files. It is sad to see that sustainable harvesting of swiftlet nests, successfully done in many countries, has not taken off in India.

In the 1970s, while going through old *JBNHS* issues during the Aligarh days, I read a paper by B.T. Phillips (originally published in 1945) on photographing the Ibisbill (*Ibidorhyncha struthersii*) in Kashmir. It triggered a longing to see the bird. The opportunity came in 2002, when I went to Sikkim, where Usha Lachungpa (who worked with me on the Florican Project for some years) showed me the bird, as she knew the area where it is found. My diary entry for 11 May 2002 reads:

> Yumthang is now a tourist destination, so by the time we reached, at about 8 o'clock, jeep-loads of tourists were being disgorged from vehicles. Most were interested in frolicking by the river, an activity that seemed to need shouting and running around, and the mandatory littering of the area. Not many were keen on the placidly grazing yaks, majestic snow-covered mountains, dark forests and the vast meadows. We were the only exceptions who were there to search for the enigmatic Ibisbill. Usha located one foraging near the cold stream. Later, when a truck came to remove stones and sand, two flew away, calling agitatedly.

Later, I searched for the species in many high-altitude rivers in Ladakh but could not find them, despite tips from Joanna van Gruisen (a wildlife documentary filmmaker and a friend) on the many sites where they could possibly be found; maybe I was not allocating sufficient time for the search. While interacting with the researchers in Kashmir University, I suggested they take up studies on the Ibisbill (the other suggested study species were the Kashmir Flycatcher and the Orange Bullfinch). Iqramul Haque, a handsome and lean young man, showed interest; I took him under my guidance in 2017. First, we collected the recent site information, and also went through the data, available in an excellent book, *Breeding Birds of Kashmir* by R.S.P. Bates and E.H.N. Lowther (published in 1952).

The Ibisbill lives in the high mountains, mostly above 1,700–4,400 m, near fast-flowing cold streams and rivers, ribboned by small boulders on both sides. This member of the shorebirds group has adapted well to such a habitat – both in behaviour and plumage. Its grey-and-white plumage, black face, a natty black breast-band, dark long legs and down-curved crimson bill perfectly camouflage it on boulder-strewn streams. Its unhurried, slow-paced foraging habit makes it almost invisible. No wonder that not many people have seen the Ibisbill, and practically no detailed research has been done on this enigmatic bird in India.

There are old records of the Ibisbill on the road to Sonamarg (Kashmir), a famous tourist area about 70 km from Srinagar. A quick three-day survey along the Sindh River by Iqram and me did not throw up any sightings, so we decided to go to Ladakh, where the Ibisbill had been seen near Leh, along the Sindh River. As we left Sonamarg early one morning, at a bend of the Sindh River near Nilgrath, we heard a call that sounded like that of a wader. A quick glance revealed our first Ibisbill – not one but five, including two juveniles. For the next five hours, we noted their behaviour, interactions and habitat-use parameters. Surprisingly, the birds were quite bold and allowed us to approach to about 50–60 m, as long as we were on the other side of the stream. A good start for the first doctoral study on this uncommon but widely distributed bird of the Asian highlands. For the next three years, Iqram worked on the species and submitted his PhD thesis in 2020. Other than two papers on the Ibisbill from China, not much was known about the bird until our study in India.

Understanding the basic ecological aspects of the bird was a challenge, as the Ibisbill is considered to be an elusive species inhabiting a specific type of habitat; it belongs to a unique family – *Ibidorhynchidae* (under the order *Charadriiformes*), which is represented by a single known species, namely, the *Ibidorhyncha struthersii* or Ibisbill. The species depends on water bodies for foraging, both during breeding and non-breeding seasons.

Our surveys of all the major rivers of the Kashmir Valley yielded good sighting records; the final stretch was carried out along the Sindh River, where the Ibisbill was found to be present at several sites, but always sparsely distributed.[43] We found this to be the case in all the rivers of Kashmir and Ladakh, and concluded that the Ibisbill has discontinuous and patchy distribution, and even in apparently suitable habitats it could be absent. The group size was typically two to five individuals during the non-breeding season (September–January), except for the autumn season when the bird showed altitudinal movement (higher to lower). During this period, congregation occurred for a short time and we sighted a group of 28 individuals, which is the highest number recorded to date.

The breeding season is annual, and runs from February/March to late July/early August. The Ibisbill appears to be monogamous and shows territorial behaviour during the breeding season. A pair occupies a particular patch year after year, and chooses a shallow depression on the ground for nest-building. Nest construction starts in the second week of February, with a slight delay at higher elevations, typically on river islands or along riverbeds. Nests are built by tossing small pebbles into the chosen depression. Being a ground-nesting bird, livestock grazing and mining may reduce reproductive success, though nest trampling and destruction of breeding grounds by haphazard extraction of boulders from the shingle beds are other threats.

Since the bird is a 'habitat specialist' (partial to high altitude, boulder-strewn rivers), it will be prone to local extinction if the threats continue to exist. Therefore, immediate action needs to be taken for the conservation of the bird, as most of the areas inhabited by the Ibisbill are non-protected.[44] Research organizations and NGOs should highlight the plight of the bird and impress it upon the concerned authorities, such as the Department of Wildlife Protection, Jammu & Kashmir, and the Department of Mining, to ensure proper protection of its breeding grounds. The nesting areas should be marked properly and appropriate care should be ensured for successful hatching. Mining needs to be kept in check at the potential sites inhabited by the bird. Tourism activities, especially in the Baltal area, need to be regulated and camping sites should be allowed only some distance away from the river. Equally important are steps that need to be taken to raise public awareness among the local communities for the conservation of the Ibisbill.

Based on our papers and other published reports, the Department of Science and Technology, New Delhi, funded a major project – 'Assessing population distribution parameters and climate-related dietary patterns of Ibisbill across elevation gradients in the Western Himalayas for conservation planning' – under the supervision of Dr Khursheed Ahmad, the head of the Division of Wildlife Sciences, SKUAST for three years, starting in 2021. The project is still ongoing and will give us an opportunity to further study the Ibisbill's ecology, habitat selection and threats. We also intend to establish an Ibisbill Conservation Group to bring together researchers from all the range countries, such as China, Nepal and Bhutan to highlight the conservation problems (and think of solutions) of this elusive bird that has largely been outside the radar of most conservationists, birdwatchers and policy-makers. I hope the Ibisbill becomes an iconic bird of the high-altitude hill streams and rivers. That may be key to its survival!

I usually encourage my colleagues and students to keep an eye on other species in their study areas. One of them, Jugal Kishor Tiwari, worked with me on BNHS's Grassland Project (funded by USFWS) in the Banni area of Kutch. A resident of Bharatpur, Rajasthan, Jugal became so fascinated by Kutch that he is now settled there. To study Banni grassland's ecology, we established a field station in a small village called Fuley, at the edge of Banni. Like other areas of Kutch, this small Muslim village of *maldharis* (cow and buffalo herders) was surrounded by *Prosopis julifora*, an invasive tree species, and Acacia trees, some quite old and gnarled. Jugal found a pair of White-winged Black Tits, now called White-naped Tit (*Parus nuchalis*), breeding there. As soon as I got his letter sharing this, I decided to visit him. The nest was built on an old tree, and we studied it until the chicks had fledged. The information collected made its way into a scientific paper, probably the first one on this species.[45] Since that pioneering study on the species, many people have studied this enigmatic tit. Jugal continued his work on it as well and found it widely, but sparsely, distributed wherever old-growth thorn forest survived in the Kutch and Banaskantha districts in Gujarat, and in the Pali, Jalore, Jaipur, Nagaur and Ajmer districts in Rajasthan. Thanks to the good protection the birds received in some sanctuaries in the last 30–40 years, many thorn trees have matured, providing nesting sites in the form of tree hollows and snags.

The White-naped Tit is an Indian endemic, and is not found in any other country. Interestingly, it occurs in two disjunct populations: northwest India and drier parts of peninsular India, thus showing discontinuous distribution. As it appears to be a non-migratory species, there is apparently no population exchange between these two groups. But, as both these populations inhabit areas undergoing loss and degradation of habitat, BLI and IUCN categorize the White-naped Tit as a Vulnerable Species.

Here, I would like to mention an interesting incident involving the bird and my old guru, which will further prove that curiosity is the mother of invention and good science. The following is what Sálim Ali observed about the roosting of the White-naped Tit in a paper he wrote[46] in 1955:

> Before I first visited Bhuj in August 1943, my cousin Humayun Abdulali gave me the 'address' of one of these tits he had been shown about six years previously, roosting at night in a particular hole in the cross-bar of a particular gate on the circular road round Bhuj Hill. He asked me half-jokingly to try and call on his friend should I have the opportunity. On doing so at sunset on August 8, I was astonished to find the tit (same individual or successor?) at home. When peeped at through a chink, the bird swayed its head and neck deliberately from side to side. In the dim light of the hole, the white cheeks and streak down the neck heightened the snake-like effect. Three evenings later I visited the roost again at the same time, caught the bird in its hole and marked it with an aluminium ring. This ringed bird was still in occupation of its roost on April 4, 1944 (i.e. 8 months later). I have no knowledge whether the gate still exists, but it would be interesting to know how much longer this bird continued to sleep there and whether the hole has since been inherited by a successor – for the original occupant can no longer be alive.

(The gate was dismantled *c*. 1950).

The Bengal Florican studies in Assam gave me many opportunities to look at other wet-grassland-dependent species, such as Manipur Bush-quail (*Perdicula manipurensis*),

Marsh Babbler (*Pellorneum palustre*), Black-breasted Parrotbill (*Paradoxornis flavirostris*), Slender-billed Babbler (*Turdoides longirostris*), Jerdon's Babbler (*Chrysomma altirostre*), Swamp Francolin (*Francolinus gularis*), Jerdon's Bushchat (*Rhodophila jerdoni*), White-throated Bushchat (*Saxicola insignis*), Bristled Grassbird (*Chaetornis striata*) and Swamp Prinia (*Prinia cinerascens*). Not much is known about the ecology of these species and their habitats even now. Fortunately, Kaziranga Tiger Reserve, Manas Tiger Reserve, Dibru-Saikhowa, Orang National Park and floodplains of the Brahmaputra still have natural grasslands. Sadly, only a few species – namely, the tiger, rhinoceros and elephant – have dominated the media, and so few pay attention to neglected grasslands or wetland species. Throughout my career, I have highlighted the neglected species and habitats, so much so that in the MoEF they would call me 'father of neglected species'. Wetlands and grasslands are the most ignored habitats in India (and elsewhere in the world). Many species of these habitats have declined to alarmingly low numbers, but no one appears to care about them. In this context, I took up a project in 2014–16 to work on the conservation of threatened grassland birds of the Brahmaputra floodplains.[19] This generated much interest in the species, with three to four researchers now working on these birds.

There are many species, particularly endemic and near-endemic, that I would have liked to study, but one life is not enough to fulfil every dream. One such bird is White-bellied Minivet (*Pericrocotus erythropygius*), which lives in dry, tropical thorn forests (though, even in its large habitat area it is not a common species). How does this bird survive in such a 'paupered' habitat while its nine conspecifics in the Indian subcontinent enjoy the bounties of forests (in the form of nectar, insects, spiders and berries) is a question that I hope will be answered by researchers in the future.

Similarly, Kashmir Flycatcher (*Ficedula subrubra*) is another fascinating species on which my student, Ashfaq Ahmad Zarri, conducted a research in Ooty,[47] but a detailed study is lacking. It breeds in a small area in the western Himalayas (Kashmir and parts of Pakistan), and winters in the southern Western Ghats and Sri Lanka. Another species in the same area is Orange Bullfinch (*Pyrrhula aurantiaca*), which too is confined to a relatively small area in the western Himalayas. I tried to get two students of Kashmir University interested in these two species, and even agreed to guide them, but after a few days of fieldwork, sadly, they both opted for easy research topics.

By sharing the stories of these feathered friends, I would like to emphasize that researchers, while keeping their focus on their study species, should, at the same time collect opportunistic data on as many species/habitats as possible, and write papers/ articles to highlight their conservation status. India is such a vast country, with extremely rich biodiversity, that even if we work on birds for 100 years, we will not be able to study all the species. Whenever I hear statements like 'duplication of data' or 'work has already been done', I laugh, as there is so much to do in our country that even if we have a BNHS or a WII in every state, it will still not be enough.

21

For the Love of Students and Science

In my long academic career spanning BNHS and AMU, I took on only 11 doctoral students, and each one of them is special to me. Some completed their PhDs 25–30 years ago but I am still in touch with them. My greatest satisfaction is that none of them left the wildlife field, and each one made an impact on wildlife conservation in India and abroad. I wish I could talk about all of them in detail, for they are all an important part of my life. But without the space to do so, I am sharing a few accounts about some of them.

Satish Kumar Sharma was my first PhD student, and so he has a distinct place in my heart. He was selected to work in the Grassland Ecology Project, funded by the USFWS through BNHS. After I left BNHS and became a PhD supervisor in AMU, Satish enrolled under me to study wolves at Nannaj (near Solapur, Maharashtra).

While working on the bustard in the Nannaj area, I had seen wolves many a time, and one alpha pair even bred there. In wolf packs, only the alpha pair breeds while other pack members help them in hunting and raising pups. I asked Satish to take up studies on this neglected animal, about whom not much is known except that the poor animal is harassed by shepherds and *pardhis* (local hunting tribe) wherever it lives. The protected area of Nannaj had given the wolves some breathing space to breed in peace; the abundant supply of blackbuck was another attraction for these meso-carnivores. During his study, I also wanted Satish to collect data on the bustards, as a continuation of an earlier BNHS research.

Satish, a Himachali, adjusted well to the hot plains of Solapur. Perhaps, two years of MSc in Jammu had trained him to tolerate the hot climate. He spent three years in Nannaj, living in the field in a small room built by the Forest Department, spending most of his time observing wolves (whenever he could locate them). His doctoral thesis on the Indian Grey Wolf (*Canis lupus pallipes*) was accepted in 1998.

After completing his thesis, he applied for a lecturer's job. At that time, a few temporary positions for lecturers were available in AMU's Department of Wildlife and Satish was appointed as a lecturer in 1996. My first student securing this appointment soon after submitting his thesis, one that was later confirmed, was the most satisfying moment of my life. Apart from being a good researcher, Satish is one of the finest human beings I have seen in my life, and he is also very popular among students and teachers. I have never seen him getting angry with anyone, nor come across anyone speaking ill about him. He is so cool that I consider him '*allah ki gaay*', an Urdu phrase for complimenting a person who is cool, friendly and peaceful (like a cow). He is also veridical to the core. Satish has had a very successful teaching career and is now the chairman of the Department of Wildlife Sciences at AMU.

He has also been an excellent guide to several doctoral students. One of them, Bilal Habib, a product of AMU, continued his work on wolves in Solapur. Bilal's thesis, which studies the carnivore's potential habitat in a GIB sanctuary, was submitted in 2007, and is another good study on the Indian Wolf. As Bilal is Satish's student, I call Bilal my 'grand student'. My 'grand student' is now a faculty member at WII.

In 2001, a young man named Koustubh Sharma from Bhopal came to BNHS to analyse bird data that he had collected from the Bhoj Tal, also named the city's (Bhopal) 'Upper Lake'. Being a student of physics, he had good knowledge of mathematics and statistics, but not much of birds. He was an amateur birdwatcher, and like many at the age of 19–20, lacked proper mentorship. His bubbling nature, restless behaviour, nervous shuffling of loose data sheets, his habit of excitedly naming a few birds and his repeated utterances that he wanted to learn about wildlife from BNHS, impressed me. I helped him organize his bird data and told him to contact me after completing his MSc. On obtaining on MSc from the Institute of Physics and Electronics, University Teaching Department, Barkatullah University, Bhopal, Koustubh joined BNHS in 2002. By that time, I had in hand a project on a neglected species called the Four-horned Antelope (*Tetracerus quadicornis*), also called the Chousingha. Endemic to India (and a small part of dry Nepal terai), no one had worked on this forest antelope. Koustubh was the perfect fit for this project; his parents were in Bhopal, and he had to work in Panna National Park not far from Bhopal.

The main reason I chose Panna as the project's field station was that my friends, Raghunandan Chundawat and Joanna van Gruisen, were working on Panna's tigers. Raghu is one of the finest wildlife researchers of India, and Joanna is like a sister to me. I placed Koustubh under their patronage. To be honest, I was extremely busy with several projects, including the establishment of the IBA-IBCN, the vulture project, Jerdon's Courser study, administration of BNHS and attending meetings of the MoEF. Though I could not give much attention to Koustubh personally, I knew he was in good hands.

Bombay Natural History Society is not a degree-giving institution, and so it has had an affiliation with the University of Mumbai since the 1960s; MSc or PhD dissertations are supervised by BNHS scientists but the degree is given by the University of Mumbai. Most students/staff who had secured their degree through the university in this manner had an academic background in biology. In Koustubh's case, the University refused to give him admission, as his master's was not in biology. A few meetings with the Registrar, Dean and Vice Chancellor helped in sorting out the problem, and I was able to convince the Vice Chancellor that in an era where interdisciplinary work was gaining prominence, the University should encourage such researchers.

Under the field supervision of Raghu and Joanna, Koustubh conducted excellent studies on the Four-horned Antelope to secure his PhD in 2007, on the 'Ecology,

Distribution and Behaviour of Four-horned Antelope (*Tetracerus quadricornis*)'. His field-observation skills were improved by Raghu and his English-writing skills by Joanna.

In 2005, I took Koustubh and Ashfaq Ahmad Zarri (my other PhD student who worked in the Nilgiris) to Kashmir and from there to Ladakh. The high mountains, the clear sky with the occasional dancing clouds, sparkling streams and flower-carpeted summer hillsides, intoxicated Koustubh – a state from which he has not yet recovered. I remember Koustubh telling me in our tent near Chushul wetlands that he wanted to work in the mountains. But to work in his favourite habitat he had to wait for a few years. From 2006 to 2009, he worked as an associate in the PEACE Institute Charitable Trust based in New Delhi, helping the Trust in designing field research, analysing data and developing GIS-based models, as part of the work the Trust was doing for the long-term ecological research and monitoring centre in Kuno Wildlife Sanctuary. Then, from 2009 to 2015, he worked as a data analyst in the Wildlife Protection Society of India. At the same time, he also analysed Raghu and Joanna's data on tigers, and developed interesting models for their work.

After a stint with the Nature Conservation Foundation (NCF), Koustubh now works in his beloved mountains as a senior regional ecologist at the Snow Leopard Trust (SLT) in the Kyrgyz Republic. At SLT, he has quickly climbed the career ladder to reach the position of the director, science and conservation. To say that I am proud of him will not be an exaggeration!

Becoming a partner of BLI brought many benefits to BNHS, in the form of joint projects, fundraising and capacity development. An interesting project that was an outcome of this partnership was a study of the elusive Jerdon's Courser. Together with the RSPB, BLI, University of Reading and Cambridge University we launched the study in 2000.

The bird was rediscovered by Bharat Bhushan of BNHS in 1986, but no systematic work had been done on the species since then. Bharat Bhushan left BNHS in 1994, and so I was looking for a young researcher who could do the fieldwork alongside me. My friend Dr Ajith Kumar told me to come to Valparai (Tamil Nadu) to meet a young man who was interested in studying nocturnal animals. Valparai also beckoned, as I was also looking for a field station for my Grassland Ecology Project in the Western Ghats. In 2000, I met P. Jegannathan (whom we called Jegan), a thin young man, who had completed his MSc in Wildlife Biology from AVC College, Mayiladuthurai, in 1998 and was working as a volunteer with Ajith.

Jegan was called to BNHS for a formal interview and before long, he was off to the Sri Lankamalleswara Wildlife Sanctuary in Andhra Pradesh to start the work on this elusive, nocturnal bird, which less than 20 people had seen until then. Jegan began searching for the Jerdon's Courser at night with Aitana, a local trapper who had caught the bird in 1986 for Bharat Bhushan. Success came on 21 September 2000 at 11.40 p.m., when a Jerdon's Courser was seen clearly in the searchlight. The next morning, Jegan called me from Siddavattam (a village close to the Sanctuary) and in a trembling, excited voice, reported the sighting of the bird. I could make out that just one sighting had hooked Jegan to this species. I was proved right when Jegan saw it a couple of times, and also began recording its calls. Prof. Rhys Green from Cambridge University and Prof. Ken Norris from the University of Reading designed a protocol to study the bird through track records and camera trapping. I must accept that I did not play any role in the study design and the full credit should go to these two professors from the prestigious universities; my only involvement was that I visited Sri Lankamalleswara a couple of times to see Jegan at work.

Jegan enrolled under me for his doctoral study on the 'Population Status, Geographical Distribution and Ecology of the Jerdon's Courser (*Rhinoptilus bitorquatus*)'. Our first project centred around the conservation of the endangered Jerdon's Courser in India, from August 2000 to March 2004, followed by another on 'Large-scale Habitat Mapping and Local Conservation Initiatives for Jerdon's Courser, India', which ran from September 2004 to July 2008; both were supported by the Darwin Initiative for the Survival of Species, UK, and guided by Profs. Rhys Green and Ken Norris. These two back-to-back projects provided us some information about the call, food and habitat requirements of the Jerdon's Courser.[48] We found that the bird prefers lightly grazed areas, with scattered bushes of 1–2 m, and occasional small trees. It does not live in thick forests and areas with tall, dense grasses. It feeds mainly on insects, worms and other invertebrates. As the habitat of the Jerdon's Courser was under threat (due to the development of a canal), in 2009 we developed a 'Jerdon's Courser Recovery Plan', with the help of the Forest Department and various organizations. It was published as an official document of the Andhra Pradesh Forest Department.[49] Based on our recommendations, a few modifications were made in the canal's design, though not as much as we wanted. For example, since the canal would be a barrier for the movement of livestock for grazing purposes (these light-grazed areas are preferred by the Jerdon's Courser), we had suggested adding a few bridges for animal crossing, a recommendation that was implemented.

In 2013–14, another project on the Jerdon's Courser began, funded by the Mohamed bin Zayed Species Conservation Fund and the RSPB. Apart from Jegan, a new research assistant was involved in the project. Unfortunately, by this time, due to the direct and indirect impact of the canal, lack of administrative interest and pressure from the local people, the main habitat of the Jerdon's Courser was hugely modified. Despite this, our researchers were able to detect the rare nocturnal bird, though infrequently.

At the end of 2009, Jegan left BNHS and joined Nature Conservation Foundation (NCF), in Mysore, where he works on nature education for school children, outreach and citizen science programmes. The shy, young man who joined BNHS in 2000 has now become an excellent communicator, and has developed innovative communication material, while also doing a fine job of distilling technical research papers into simple language for the general public. Jegan has written hundreds of articles in newspapers and magazines in English and Tamil, reaching a large audience with his conservation messages. People like him stand proof to the fact that India has come a long way in wildlife research and nature education over the last three to four decades.

It is difficult to describe Rajat Bhargava: childlike, innocent, garrulous, irritating, lovable, sincere, cheerful, voluble, obstinate, obedient, pig-headed, diligent, creative, determined...so many words come to mind. Some of the above-mentioned adjectives may read contrary to each other, but they describe Rajat's complex personality. For instance, he is affectionate towards his parents, brother, sister and family, but will disappear for six months without telling his anxious parents about his whereabouts. He can be obstinate, but can listen and change his views if convinced properly.

I taught Rajat in AMU in the early 1990s. As he belongs to a strict Brahmin family, his parents were reluctant to send their son to a 'Muslim University', fearing that he would be 'spoilt', and turn non-vegetarian! When Rajat told me about the apprehensions of his family, I made sure that he was welcomed and made comfortable in the Department, and that his culinary sentiments were respected.

Rajat came to know about the MSc post-graduate course from J.C. Daniel's wonderful article on Sálim Ali in the *Reader's Digest*, after the latter's death in 1987. Rajat, then a fresh graduate, wrote to J.C. Daniel about a career in ornithology. Daniel's advice (that came enclosed in the prompt reply) was to opt for admission in either WII or AMU. As WII used to admit students for the MSc programme only every alternate year, Rajat's option was AMU. Prof. Musavi was impressed with Rajat's interview, despite the fact that Rajat did not do well in the written test. After the interview, Prof. Musavi told me (I was then in the first year of my teaching at the Centre of Wildlife) about a young student from Meerut whose knowledge of birds was incredible. One meeting with Rajat, and Musavi's assessment was proved right. The day Rajat joined, I had planned to take a written test again. He failed in the test and got 3 out of 10 marks, while the more successful students secured 6 or 7 marks. However, Rajat answered all the three bird questions correctly, which no one else had, as those topics were not taught in the typical BSc course. This impressed me further. My interest is always in students who are interested in wildlife, not the textbook nerds with high marks.

I began admiring Rajat for his amazing knowledge, not just of birds but also of bird trapping. He was not even 20 years old then but his knowledge of bird trade, bird keeping, trappers' modus operandi, types of cages and places where birds are caught, was encyclopaedic. He was also fond of telling stories of his experience with trappers, to anyone who had the capacity to tolerate his volubility! I found his stories fascinating, and they reminded me of my experiences with Ali Hussain, the famous bird trapper of BNHS.

Meerut being a four-hour drive from Aligarh, Rajat would disappear there frequently. Once, he did not attend classes for two days. I wrote to his father, Vishu Bhargava, a famous advocate in Meerut, as I was fond of Rajat and wanted him to do his MSc sincerely. Within three days, his father came to Aligarh along with my elder brother's friend (my brother had worked in Meerut for a few years, and had friends in common with the Bhargavas) and his recalcitrant son. The senior Bhargava brought along my brother's friend so he could offer *seefarish* (recommendation/support), I presume. A joint reprimand was issued to Rajat and all was well after that. Soon, he became the best student in the class. Later, Rajat confessed to me that he frequently went to Meerut to feed his caged birds, as no family member looked after them!

After that I started giving him leave whenever required, as the issue at hand – the life of birds – was important to both of us.

Rajat's large house was located near Meerut's famous Begum Bridge, named after Begum Samru, an eighteenth-century *femme audacieuse*, who became a Catholic Christian and married a mercenary, Walter Reinhardt Sombre. Adjacent to Rajat's house was a colony of animal and bird traders, so Rajat had grown up listening to the chatter of newly caught birds. From his second-floor window, he could see the *baheliya* trappers, bird cages and all the dirty games that went on within this illegal trade. From the bird trappers and traders, he learnt about keeping and caring for birds, the size of cages required for each species, their upkeep, species-specific feeds and traditional methods of treating sick birds. His childhood experiences helped him in his later life. He told me that 'my life coaches were my backyard neighbours'. Before the blanket ban on bird trade, including export, in 1990–91, Meerut used to be one of India's biggest bird trade and export centres, second only to Kolkata. Rajat enjoyed the privilege of observing more than 200 species of birds (which were exported from Meerut), including several threatened and rare species. As a child, he learnt to recognize most birds by their local trade names.

In the then Centre of Wildlife (now Department of Wildlife Sciences), the MSc dissertation fieldwork was typically done in the vicinity of Aligarh. But when Rajat told me that he wanted to do the work in four districts of UP, for which he got funding from WWF-India, I allowed him to and convinced the then chairman, Prof. Musavi, that Rajat should be allowed to go to the field on weekends. When I rejoined BNHS in May 1997, I helped him secure admission for a PhD from the University of Mumbai. However, keeping an eye on him was a difficult task, as he was travelling all over India, studying the bird trade. Finally, he submitted his thesis, a masterpiece on bird trade, in 2012.

Rajat is an extraordinary storehouse of knowledge when it comes to the subject of bird trade. While working on the Bengal Florican in the 1980s, I became interested in another grassland-obligate species – the Swamp Francolin (*Francolinus gularis*). At that time, habitat destruction and shooting were considered the nemesis of this species, but Rajat told me that its trade was also another major reason for its decline. He showed me, as an example, four Swamp Francolins kept in small cages in the so-called Van Chetna Kendra (Forest Awareness Centre) in Bulandshahr, a town in UP. Later, Rajat also told me about the trade in Green Munia (*Amandava formosa*), Finn's Weaver (*Ploceous megarhynchus*), Intermediate Parakeet (*Psittacula intermedia*; now no longer considered a full species) and other birds. His knowledge of bird trade was a great help in my lectures, meetings with officials and when I was writing the book, *Threatened Birds of India*, in 2012.

We conducted many surveys together, despite his annoying talkative habit. Imagine tolerating constant chatter for weeks during surveys! No wonder he is called '*chhota tota*' (small parakeet) by his trapper friends. During meetings with experts or forest officials, he would often embarrass me by interjecting in the middle of my, or my guest's, sentence. His impulsive behaviour led to both interesting and embarrassing situations. During a survey in the Thar Desert in 1994, for example, we were in a village called Nachana, near the Indira Gandhi Canal. We had gone to a small *dhaba* for food, and I told the owner that we (myself, Mehboob and Rajat) were 'pandits' and needed pure vegetarian food in clean plates. My 'pandit trick' was mainly to ensure cleanliness in service. Rajat, who had gone to wash his hands, had not heard my instructions. When he returned from the washroom, he asked the owner, 'Do you have boiled eggs?' The owner was shocked. He asked in dismay, '*Aap kaise pandit hain jo anda khate hain* (What kind of pandits are you who eat eggs)?' Before I could answer, Rajat, as always, blurted out, '*Yeh dono to sab khate hai, main hi hoon jo bas anda khaa sakta hun, mass nahi* (They both eat meat, I can only eat egg, not meat).' I have never been so embarrassed in my life as I was then! We got the vegetarian food, accompanied by dirty glances from the owner. On my subsequent visits, I could never go back to that particular *dhaba*.

To feed his restless intelligence, I encouraged Rajat to take up new studies. When he said that he wanted to understand the origin of Rothschild's Parakeet or Intermediate Parakeet (whose specimens were shipped from Bombay to the American Museum of Natural History in 1895), or its whereabouts (collected presumably from Sikkim), I told him to go ahead. This was in 1997. There are seven specimens of this parakeet, which have never before been seen in the wild, in The Rothschild Bird Collection of the American Museum of Natural History. Rothschild's Parakeet's taxonomy had always been debated and the question was whether it was a hybrid or a full species. We needed to conduct genetic studies to answer this. Unfortunately, we could not get funding for DNA analysis, but in 1999 Dr Pamela Rasmussen of the Smithsonian Museum, and Dr Nigel Collar of BLI proved that it was a hybrid between Plum-headed (*Psittacula cyanocepha*) and Rose-ringed (*Psittacula krameri*), but mostly between Plum-headed and Himalayan Parakeet (*Psittacula himalayana*).[50] Rajat was heart-broken, as his quest was not answered due to lack of funds. However, I encouraged him to shift his focus to other species, such as Finn's Baya. The bird was considered lost till Dr Sálim Ali discovered[51] it in the Kumaon terai in 1957. Dr Sálim Ali's first student, V.C. Ambedkar, conducted extensive studies on the bird and published papers in the *JBNHS*. I was delighted that 40 years after Dr Ali's discovery, my student was going to study the bird.

As Rajat got good information from his trapper gurus on the Finn's Baya, I sanctioned a small grant from the IBCN in 2002 to conduct a rapid survey. Rajat's survey reflected a shocking decline of Finn's Baya populations, as most of the bird's wet grassland habitat had been destroyed in the last 40 years. Much later, in 2016, BNHS sanctioned funds from the Sálim Ali Nature Conservation Fund (SANCF) to revisit the habitat areas, and to my horror much more shocking results came out from the fieldwork. The species is now on the verge of extinction! Rajat's report 'Status of Finn's Weaver in India: Past and Present', resulted in BirdLife-IUCN classifying the Finn's Weaver under the 'Endangered' category in 2021. It also led the Government of UP to fund a conservation breeding project for the bird in the Hastinapur Wildlife Sanctuary, which is still awaiting Central Zoo Authority approval in 2024!

During his stint at WWF-India, Rajat kept an eye on many neglected species, such as Green Munia, a bird endemic to India. He informed me that it was in high demand across the world (like other munias). Considerable numbers were exported from India until 1990–91, when the Indian government banned all bird export. To cut the story short, thanks to the efforts of BNHS and Rajat, a conservation breeding programme was started in Udaipur by the state Forest Department under the supervision of Rajat. As I write this in

February 2024, I am happy to report that successful breeding of the Green Munia has taken place in captivity.

Rajat is unmarried. His parents sometimes blame me for his personal choice! Rajat has totally devoted his life to study the bird trade, before it dies out completely. Rajat's dream is to create a museum on 'Cages Through the Ages', showcasing traditional bird cages. However, the traditional art is dying, as the bird trade of domestic birds is disappearing (and rightly so). His other dream is to write a book titled *Chidiyawalas – The Indian Bird People*, about bird trappers and keepers. I wish to see his dream fulfilled.

While surveying states for the GIB and the two floricans (Lesser and Bengal), I collated data on many other species, including storks, such as the Black-necked Stork, Painted Stork, and Lesser (*Leptoptilos javanicus*) and Greater adjutants. At that time, not much attention was given to storks in India, except for some work on the Greater Adjutant by Prof. P.C. Bhattacharya from Guwahati University and his students. The USFWS showed interest in funding further studies, which resulted in the approval of the 'Stork Ecology Project', that was executed through AMU from 1994 onwards. We recruited three researchers for this project; one of them was Farah Ishtiaq. I asked her to work in Keoladeo National Park, to which she agreed without batting an eyelid. Farah's excellent work on the comparative ecology and behaviour of endangered storks (Painted, Black-necked, White-necked and Openbill) resulted in her PhD in 1998. She also did wonderful studies on the Forest Owlet (*Heteroglaus blewitti*) in a forest patch in Shahada, Maharashtra. Later, Farah shifted her focus of research from bird species to bird diseases and genetics, and went to the US and UK for many years. Currently, she is working as a principal scientist, Tata Institute for Genetics and Society, GKVK Campus, Bengaluru, where she is studying the epidemiology of avian malaria and spread of the disease in high-altitude malaria-free zones in the Himalayas. Another recruit to the stork project, Hilloljyoti Singha, completed his PhD under me in 1999. Hillol lost his father, Tapan Kumar Singha, when he was 10 years old, and his brave Bodo mother raised her two sons singlehandedly. Despite being 54 years old, Hillol is still a young, energetic and excitable man, willing to fight for his strong views on social issues. Due to his restless nature, no-nonsense attitude and decadence of our university systems, he changed five jobs after completing his PhD, and is now a professor in the Department of Zoology, Bodoland University, Kokrajhar, Assam. The third person to get a PhD under the stork project was Gopinath Maheshwaran. Today, he is a PhD guide at the University of Calcutta, and has also guided research on the extremely rare White-bellied Heron (*Ardea insignis*) in Arunachal Pradesh, perhaps the first study on this bird. Maheshwaran is scientist in-charge at the Ornithology section of the Zoological Survey of India.

Girish Avinash Jathar was my eighth student, who worked on the Forest Owlet (*Heteroglaux blewitti*) for his doctoral work. Girish is presently deputy director of an NGO, Srushti Conservation Foundation, based in Pune. He looks after the research wing of Srushti, whose main focus is climate change, human–wildlife interactions and sustainability. Another PhD student was Sujit Shivaji Narwade, who studied the birds of the Deccan for his doctorate, in particular, the GIB. At present,

Sujit is deputy director at BNHS, and works on the Lesser Florican and the GIB, continuing my work on these species.

Someone who chose to work far away from home for his doctoral studies was Ashfaq Ahmad Zarri, who studied the Nilgiri Laughingthrush for his PhD. When he saw the Kashmir Flycatcher (*Ficedula subrubra*) wintering in the Nilgiris, he also conducted research on this little-studied bird. Earlier, it was thought that the Kashmir Flycatcher passed through the Western Ghats to winter in Sri Lanka, but Ashfaq found that some individuals stay through the winter in the Nilgiris. He also found that even in those winter quarters, they live in pairs, which is rather unusual for flycatchers. Presently, he is the registrar of Kashmir University, Srinagar.

I list Usha Lachungpa as my PhD student, though for many personal reasons she could not submit her PhD thesis. She is one of the greatest conservationists and field naturalists of India. She worked in the Florican Project but later joined the Sikkim Forest Department in 1986, as a project officer in the Forests, Environment & Wildlife Management Department of the Government of Sikkim in Gangtok. She went on to become a research officer, before finally retiring after 31 years of service as the principal chief research officer and additional director, Sikkim Biodiversity Board. Together, we conducted many surveys in Sikkim to study birds.

When people ask me about the most satisfying part of my life, my mind leaps to my students, such as Sujit, Satish, Jegannathan, Ashfaq, Koustubh, Rajat, Hillol, Maheswaran, Farah and Girish – they are my real *sarmaya* (Hindi word for capital, or principal amount) in life.

22

My BNHS Colleagues

My BNHS colleagues have a special place in my life, and they have had a significant influence on me and my day-to-day life. In a career spanning four decades in BNHS, I had the privilege of working with many honest, sincere and wonderful colleagues. I wish I could write about them in detail, but space constraints make that impossible.

To name a few, Dr Gayatri Ugra, a top-class editor who taught me the art of editing; Vibhuti Dedhia and Sonali Vadhavkar from the Publications Department are the other colleagues from whom I learnt many tricks of the editing trade; Ruby Madan and Divyesh Parikh from the Central Marketing Department taught me the intricacies of marketing a product; Varsha D. Chalke from the Accounts Department taught me how to read a balance sheet; Nirmala Barure brought incremental improvement to the BNHS library and also developed the archives. Other noteworthy colleagues include Isaac David Kehimkar, a multifaceted personality who was a good PR manager, writer and a butterfly expert; Swapna Prabhu, plant ecologist and taxonomist; the late

M.G. Mathews, our administrative officer; Vasant Naik, Sunil Ghavnalkar, Santosh Mhapsekar, Shantaram Karambele and Swapnil – from the administration team – and finally, Sadanand Shirsat from the library. I apologize if some names have been left out inadvertently.

On the conservation front, I had colleagues like Vibhu Prakash (whom I have known since 1981), who is an ornithologist and the 'father' of vulture conservation in India; S. Balachandran, a top-class ornithologist and bird ringer; Ranjit Manakadan, who worked with me on the bustard project; Bharat Bhushan, now honorary secretary of BNHS, who worked with me on the bustard project and rediscovered the Jerdon's Courser in 1986; Vithoba Hegde, a wonderful staff member from the BNHS Museum; Zafarul Islam from the initial years of the IBCN-IBA programme, from 1999 to 2004, and with whom I wrote three major books; Raju Kasambe, a fine ornithologist, writer and speaker, who ran the IBA-IBCN programme until I retired; and Prashant Mahajan who ran the CEC and later the IBA-IBCN programme for a few years. Among the younger crop of scientists who worked with me, there are too many to name.

Since I have always promoted women's involvement in all walks of life, including conservation, I will write here about two stellar young women with whom I had the pleasure of working as director.

~

Dr V. Shubhalaxmi was a product of the Conservation Education Project (CEP) of BNHS and the RSPB. A key CEP initiative was the setting up of Conservation Education Centres

(CECs). She joined CEP as an administrative assistant in 1993, but within four years, she rose to become Mumbai CEC's education officer, a role she held until 2006 (the Delhi CEC also came under her eventually). Once the CEP ended, and the building and activities were properly established, BNHS converted CEP in to Conservation Education Centre. Given her skills, she was selected for a Fulbright Fellowship under the Indo-American Environmental Leadership Programme in 2002. When Prashant left BNHS in 2006, our first choice for running the CEC programme was Shubhalaxmi, as she had acquired all the abilities required for this profile. Our selection was validated when she took the CEC cell to new heights, by starting innovative programmes such as 'Breakfast with Butterflies', 'Monsoon Magic' and many courses, such as Basic Course in Entomology, Basic Course in Ornithology, Basic Course in Herpetology and Leadership Course in Biodiversity Conservation. These courses, all meant for amateurs, were extremely popular among members and corporates.

Shubhalaxmi's interests were varied but her focus was butterflies and moths. She did her PhD on 'Ecology of Moths of Sanjay Gandhi National Park (SGNP), with Special Reference to *Saturniidae* and *Sphingidae* families'. In India, many people acquire PhDs but few can write books. Shubhalaxmi wrote many, such as *Field Guide to Indian Moths*, *Green Guide for Teachers* and *In Harmony with Nature – A Teacher's Handbook on Learning for Sustainable Living in Maharashtra* (with Prashant Mahajan). Shubhalaxmi went on to conduct 42 courses for amateurs and professionals, developed five nature-based mobile applications and eight online courses under iNaturewatch Foundation (which she established after leaving BNHS). She went on to complete as many as 55 external projects, including the development of butterfly parks for corporate houses.

Anticipating a leadership change in BNHS, Shubhalaxmi left the organization in 2014, ahead of my retirement, and went on to found the Ladybird Environmental Consultancy in 2014, and later iNaturewatch Foundation, with Isaac Kehimkar, in 2016. Staff like Shubhalaxmi, who are involved in the day-to-day running of the organization and execution of internal decisions, made my directorship at BNHS successful.

When Neha Sinha joined BNHS as head, Conservation and Policy, in 2012, she brought a fresh youthful air, style and glamour to BNHS, much like Ravi Sankaran had in 1985. I believe that good organizations should have a mix of brashness of youngsters, sedateness of the experienced and wisdom of the grey-heads. I was highly impressed when we had interviewed Neha – who wouldn't be on finding out that she was the winner of the INLAKS Foundation Scholarship to study MSc (Biodiversity Conservation) at the Oxford University in 2010–2011? She also went on to win the Graduate Travel Fund at Mansfield College, Oxford University, for her research dissertation on tiger conservation during 2011 in the Sariska Tiger Reserve, Rajasthan. By then, she had already written more than 50 articles in leading newspapers and magazines.

While working for BNHS, she operated out of Delhi, so I would meet her during my frequent visits to the capital for official meetings. I found that she was at ease in the corridors of power as much as she was while studying birds, her prime interest. Not a qualified biologist in the true sense, she gained natural history knowledge through reading, interacting, writing and travelling to the wilderness.

Our first field visit together was to Nagaland, in November 2012, to a little-known village called Pangti in Wokha district, where vast numbers of Amur Falcons (*Falco amuresis*) congregate for a few weeks on migration. Conservationist Ramki Srinivasan, who unfortunately died at the age of 49 in December 2022, had widely publicized the mass killing of Amur Falcons by local hunters in October 2012. Neha was in touch with Bano Haralu (journalist, environmental activist and a gritty young woman), who was then working with Ramki, the state government and other organizations to protect the falcons.

After meeting Bano at Dimapur, we commenced a six-hour nerve-racking journey to Wokha, on a narrow, potholed hilly road. Neha had just recovered from a road accident, so I was worried about her, but she kept her composure, hiding the pain of her newly healed bones. Meeting the village elders amid the tense atmosphere of hostility in the village of Pangti, which had earned notoriety due to the mass killing of falcons, highlighted Neha's keen negotiating skills. In the following days, even as we continued our meetings with the village elders, we kept thinking about how we could implement a conservation plan in an area that had never seen or heard of bird preservation. After two days of dawn-to-dusk interactions, we decided that training and educating the locals, particularly the youth, would be the best option to stop the mass killing of falcons. The Amur Falcon would go on to become the symbol of an overall environmental awareness campaign. Our field reports to BNHS, the RSPB and BLI, based on this visit, resulted in a three-year project on environmental awareness, with Neha as its leader.

Neha and I went on to work on multiple projects together, touring many areas, meeting ministers, bureaucrats, army officers, village leaders and community leaders. There are many points of convergence in our beliefs – for instance, we espoused that wetlands are not wastelands during our articles and lectures. We also read and cheer for each other's books and columns. I have always appreciated her articles, and was delighted (and proud) to read her book *Wild and Wilful*, which came out two years ago.

After my retirement in 2015, she continued her good work, but the new management at BNHS could not understand and appreciate her type of advocacy and policy work – the work that brings subtle changes and cannot be easily quantified. One EC member even questioned her writing popular articles, accusing her of 'writing for money'. The quality that I appreciated most in her was questioned by a person who had never written any article himself! If a Policy and Conservation Officer of BNHS does not meet people, negotiate with decision-makers, does not write articles to reach a wider audience, then what is she supposed to do? Finally, in 2023, Neha left BNHS and was soon inducted by WWF-India as head, Policy and Communications. More power to her pen and voice!

23

Field Assistants

The book, *More than Just Footnotes: Field Assistants in Wildlife Research and Conservation*, edited by Prof. Ambika Aiyadurai and Ms Mamata Pandya, which I released in IIT-Gandhinagar on 29 August 2023, highlighted the importance of field assistants. Being leftist in ideology, and democratic and liberal to the core, I always gave importance to my field assistants. For me, every human being is equal, no matter what his/her caste, religion, profession, sexual orientation, colour, creed or economic status. At my field stations, or even when I went out for an official workshop/event with staff, I would insist that we all eat together, stay together and sometimes even share a room. This treatment had an interesting result – my field assistants felt more like 'equals', but at the same time, I sensed them thinking that I was a little bit different from other *sahabs* or bosses. That was fine with me, as long as they felt comfortable and were respected for the work they were doing with me.

I had the opportunity to work with amazing field assistants, starting with Munna aka Parmar Singh in Karera in 1982. Munna belonged to a poor Thakur family of Fatehpur village, and owned a small piece of rocky land on which no crops grew. Unlike most Thakurs, Munna was a diminutive man of 5 feet, without the signature moustaches of his fellow brethren, and lacking the arrogant gait of the so-called 'upper castes'. While I was busy studying bustards, Munna would cook food for me, the quality of which would shame even the worst roadside *dhaba*. Even after six years of working with me, the cooking skills and food quality were the same. In hindsight, I admire him for being consistent in his poor culinary skills!

Munna was honest, hard-working and a consummate consumer of gossip, some of which he would relay to me. In reciprocity, he would tell villagers about my work, BNHS, my guests and assistants. I always carried a bag in the field, crammed with a notebook, two extra pens, a packet of biscuits, my old camera, lens and binoculars; I would not allow anyone to carry it for me. Soon, the news spread that I carried a hidden pistol in the bag, which was why I was not afraid to go anywhere, anytime, including midnight returns from the field (though this was mainly due to the breakdown of the old vehicle). I would also spend nights, especially in winter, in the 'gypsy' hut near Fatehpur village. As the area was located in the badlands of Gwalior, Bhind, Morena and Shivpuri – infamous for dacoity and looting at that time – some people wanted to steal my 'pistol'. Luckily, on the 'D-day', by chance I went to Karera to stay overnight and finish some official work, totally oblivious to the threat that loomed purely on a false assumption. Much later, when he found out, Munna told me that people took my brief overnight move to Karera and the prescience of the impending threat that day as proof of my superpower. I requested Munna to keep the myth alive, so I could go anywhere at any time.

Munna worked for me for six to seven years. Even when there was no one in the field station at Fatehpur, we paid him a salary; this was the least we could do for this poor man. Before leaving Karera–Fatehpur, I wanted to ensure that Munna was employed by the Forest Department. But by 1990, the bustard population had declined, and the Forest Department lost interest in the Sanctuary, leading to curtailment of funds. After leaving Karera, I kept in touch with Munna for almost two decades, but I lost contact eventually. I hope he is doing well.

When I was looking for a good field assistant for my Bengal Florican project in 2014, Bridesh Singh's name was recommended, as he had worked for a WTI project for two years. I soon realized what a wonderful asset he was in the field and at the field station. Having studied till grade 12, Bridesh could independently take field data. He was also a friendly, easy-going person, who got along well with Rohit Jha and Rahul Talegaonkar, the two researchers on the project. The four of us lived in a two-room apartment adjoining a movie theatre in Pallia, a town in UP's Lakhimpur Kheri district. The main reason for living in the compound of a movie theatre was the regular supply of electricity during the four daily film shows. Pallia was notorious for irregular electricity supply, so the move ensured that we would get at least 10–12 hours of electricity from the large generator of the movie theatre. It was another matter that the theatre showed unwatchable movies with *dishum-dishum*, double-meaning dialogues, gore and sex. None of my young staff showed any interest in entering the theatre, although the owner, a friendly man, offered to give us free access. When the Bengal Florican project ended in 2016, Bridesh started working in my Sarus project, which continued for another three years. He shone in this role – most of the field data was collected by him, along with Suhail Ahmad and Fazlur Rahman from Mohammadi town. Under this project, we surveyed seven districts of UP and Bridesh accompanied me on all my field surveys. His name appeared in the final report, and also on the paper that I wrote later on Sarus Crane.[52] This was the least I could do for him. I am happy to write that Bridesh is now settled in his village, Bhagwant Nagar, and looks after a government ration shop. Nothing gives me more pleasure than to know that my assistants and students are all well settled in life.

I first met Mehboob Alam in November 1980, when he accompanied his famous father, Ali Hussain, to Keoladeo National Park, as a newly minted 'adult' at age 14. He would join his father at night, to trap birds for BNHS's Bird Migration Project. I was not happy that an underage boy was working, particularly in the night, but I had no control over the project. At Bharatpur, we worked with two types of bird trappers: Mirshikaris who were Muslims, and Sahnis who were Hindus. Both groups had different ways of catching birds. Sahnis were mostly fishermen and opportunistically caught birds by erecting nets just above the water, and chasing the birds towards the nets to trap them, while Mirshikar bird trappers used many types of devices, such as nets, nooses, clap traps, *lahsa* (a type of gum) and throw-nets. Mirshikaris also caught the birds alive, without causing harm or injury, as in Islam, eating dead animals or birds is prohibited – they have to be ritualistically 'halal-ed' before eating. During the British period, the Mirshikaris sold feathers of egrets and herons, which were in demand for decorating hats and caps. Additionally, they sold birds as pets and also to the zoos.

Once, when I saw Mehboob and Golu Sahni, another trapper, breaking dry branches from trees, I told them the importance of such branches and twigs for the nesting of birds. Even after 40 years, Mehboob remembered this, and he once told me, 'Sir, this was your first lesson to me, to respect nature.'

In 1985, Mehboob came to Karera with his father, when the MP Government gave us permission to put colour rings on four bustards for research purposes. After studying the movement and daily rhythm of the bustards, Ali Hussain and Mehboob set up nooses at the spots that were used regularly by the wary bustards, and successfully caught two male and two female bustards. Later, the same father–son team caught Lesser Floricans in Sailana. There, Mehboob requested me to teach him driving, which I did, and soon he was driving my 'Old Girl', the Willys Jeep that I had been using since 1981.

The long trips from Bombay to Sailana to Karera, intensive fieldwork, riveting stories by Ali Hussain in the dark and the electricity-deprived nights in the crumbling field station (in Sailana) form some of my most cherished nostalgic memories from the field. From 1985, I took Mehboob under my umbrella, because of his hard work, sincerity and friendliness. Like with other assistants, I treated him as an equal, sharing meals, rooms and other resources. Incidentally, driving (which I taught him) is now his livelihood. Mehboob worked with me in five to six major projects, spanning species such as bustards, floricans and storks, and in habitats such as the Thar, grasslands and wetlands. When I began to travel extensively in my old jalopy to the remote areas of MP, Rajasthan, Gujarat, Maharashtra and Andhra Pradesh, he became indispensable. When I shifted to AMU in 1991, Mehboob came along with me. Between 1992 and 1996, we did five major surveys of the Thar Desert, covering Rajasthan and Gujarat, two surveys in Sikkim and many shorter surveys under the Stork Project. Every two or three weeks I would go to Bharatpur, where Farah was conducting her studies. Mehboob would enjoy birdwatching as much as we did.

In 1996, I was heading a project to identify important grasslands under the larger Biodiversity Conservation Prioritization Project (BCPP) run by Shekhar Singh, a professor in the Indian Institute of Public Administration, New Delhi. Due to administrative difficulties of routing the finances through AMU, I got this project through BNHS. When I shifted back to BNHS in May 1997 as director, Mehboob and Zafarul Islam, both working in the BCPP project, also came with me. As I was extremely busy in 1997–98, establishing myself as the director of BNHS, I asked Zafar and Mehboob to survey the grasslands of the Thar Desert, which they did successfully. The data was used in our final report, which was submitted to Shekhar Singh.

Mehboob was much more than a driver – he was also a top-class birder, field naturalist, photographer and a companion. After working the full day in the blistering heat or the biting cold of the Thar Desert (which can go from 52°C on summer days to 1°C during winter nights), he would prepare food, while I wrote field notes in the dim light of our petromax, in some forest *chowki* of the Desert National Park. We had a lovely time in the field for 25 years, travelling together across the country – from the remote villages of the Thar to the cold desert of the Sikkim trans-Himalaya. Two incidents remind me of the 'adventures' we faced together in the field.

After the sad extinction of bustards in Karera in 1991, a few bustards were spotted in a village called Surajpur, some 25 km from the Karera Sanctuary. To save their skins, the Forest Department officials started saying that the bustards of Karera had shifted there. This was not true, as six to seven bustards had traditionally occupied the rocky, uncultivated grazing land around Surajpur, Bhimbada, Ajaipur and Simiria villages. Vishal Singh, a competent forest guard posted in Fatehpur, went to Surajpur many times, and even saw a displaying male. I decided to explore the area in the early 1990s, during my visit to Karera–Fatehpur. On a very hot day in June, I went to Surajpur along with Mehboob. We spent three hours under the shade of a large Peepal (*Ficus religiosa*) tree, talking to villagers. When the temperature became tolerable, we went to the field and found evidence of bustards. In the evening, not finding any place to stay overnight, we decided to leave for Ghatigaon, where a forest rest house is located. The villagers showed us the route but also advised us to go at once, as dacoits were active in the forests through which the road passed. It was a dark moonless night but we started nonetheless. Soon the road became narrow and hilly, with thick forests on both sides. In the 30 km of that winding, hilly road, just one vehicle crossed us. As we climbed a hill, we saw a tall, fair-skinned man with blue eyes, dressed all in white, standing in the middle of the road, almost shining in the light of the jeep's headlight, in stark contrast to the darkness all around. Quickly locking the vehicle doors, I told Mehboob to drive fast and not stop. I pulled on my large Stetson, which I have often used to impress villagers during tricky situations (in rural areas, not many people have hats and those wearing one, especially an official-looking one, are considered important). The 'white' man kept on staring at us; probably he was trying to figure out the type of vehicle that was hurtling towards him. With great dexterity, Mehboob swerved the Gypsy away from the figure and quickly went past him. The rear mirror showed that he was still standing in the middle of the road, but there was no looking back for us. We drove on, stunned by the mystifying white apparition that had appeared on the road. For the first time in my life, I was frightened. We calmed down only when we started hearing the trucks plying on Gwalior–Shivpuri road. In a trembling voice, Mehboob asked, '*Sahab, woh kya tha* (What was that)?' That broke the deathly silence between us. For the first time, we took pleasure in the truck noises and their loud honking on the road. Not finding a room in the guest house at Ghatigaon, we stayed in a *dhaba*, anchoring ourselves to the noisiest and busiest spot within it to feel safe. Sleep eluded us that night.

The next morning came the news that a bus had been looted that night on the very route we had taken. When we told locals that we had driven on the same road the previous night, they said that we were lucky to be driving a green Gypsy, a vehicle generally used by police in that area. Hearing that, things became clear. The 'white apparition' had been checking the incoming vehicles, and trying to figure out who was inside them. As soon as he realized that it was a green Gypsy, with a person in front wearing a hat, he had assumed we were policemen, and so did not attempt anything. In all likelihood, his accomplices had been hiding nearby, waiting for a signal to pounce on us. Our green Gypsy had saved us.

The second incident wasn't as dangerous. In June 1994, we were in Upper Sikkim with Usha Lachungpa. While leaving Gangtok, Usha bought some green vegetables, mutton and pork for her friend, Chingya Bhutia, one of the last shepherds of Tibetan origin living in North Sikkim. After making the long, 130 km journey from Gangtok to Mangan, Chungthang and Lachen, we stayed in a small hotel in Thangu, run by a young Rajasthani man married to a Sikkimese girl. Thangu is at 4,000 m and it was once a major stop for Tibetan traders and shepherds travelling to the erstwhile princely state of Sikkim. Now, it serves as the main town for tourists going to Yumesamdong (4,663 m) or Zero Point, and Gurudongmar lake and monastery (5,340 m). The lake is named after Guru Padmasambhava or Guru Rinpoche, the founder of Tibetan Buddhism.

The next morning, we trekked to the hillside where Usha's friend, Chingya, had an isolated hut, where he had lived for six months with his horses, yaks, dzo (a hybrid between yak and cattle) and sheep. Usha had promised to show me the Tibetan Snowcock (*Tetraogallus tibetanus*). As the journey was long and steep, Usha had arranged a yak for me, but I found it difficult and preferred to walk. We reached the hut by 3 p.m. The hillside was east facing, and the sun was already behind the high mountains. Chingya greeted us warmly and shepherded us inside his hut, where a fire was raging in the hearth. All around, household items were arranged neatly, including the thick carpets made of yak wool, on which we were asked to sit. The warmth from the fire soon dissipated the tiredness, at least momentarily. Usha gave him the vegetables, mutton and pork that we had brought from Gangtok. I have never seen anyone holding *baigan* (brinjal) as reverentially as Chingya did. At that moment, I realized the value of things we take for granted, especially things that are not easily available to the remote communities. He offered to cook vegetables for us, which we refused, as they were for him and were intended to serve him for many days. Instead, I asked him to cook yak meat for us. Many nomad communities have a traditional method of preserving meat, particularly in the cold climates. A large chunk of meat is hung above a slow-burning fire, which dries the meat without cooking it. Smoke inhibits decay and is a deterrent to insects and bacteria. It also brings a smoky or woodsy aroma to the meat.

Even before sunset, *chang*, a ragi-based alcohol, started making the rounds. Usha made sure that my glass was never empty, as she was worried that I would develop breathing issues due to the cold and elevation. She said that the locally brewed *chang* would keep me warm. The yak had been slaughtered six months ago, but the meat was tender and delicious, despite the limited ingredients that Chingya had. Mehboob refused to eat as it was not *halal*; the poor fellow had to survive on boiled Maggi. *Chang* was also not kosher to Mehboob. By 9 p.m., it was time to go to sleep. While others slept soundly, my bladder did not allow me to sleep properly the whole night. The *chang*, instead of making me warm and sleepy, had a diuretic effect. I had to step out four times on that dark, cold night, walking 150–200 m from the hut to relieve myself. As the hut had become quite warm due to the fire, we had removed many of our clothing layers, but each time I had to go out, it involved pulling on three or four layers, as the wind outside was bone-chilling. A night to remember, but not perhaps for the happiest of reasons!

The night's discomfort and the tiredness of travelling to the distant place were forgotten when Tibetan Snowcocks began calling early in the morning. They were literally spread across the whole mountain face, calling, fighting and chasing each other. It was as though a curtain had been raised for a rumbunctious on-stage tango. The avian drama went on for an hour, and as the sun rose the snowcocks decided to forage silently, merging perfectly into the low bushes and boulder-strewn hills, the rivalry forgotten quickly. On returning to Gangtok, I decided to buy something more nourishing than Maggi for Mehboob, in case we came across more non-*halal* meat. After all, he was my best companion in the field, so I had to look after him well. For me, nothing is off limits, be it a crocodile in the famous Kenyan restaurant, *Carnivora*, ham *salami* in Germany or frog legs in Taiwan. I eat anything as long as it is legal in that country.

Once, while surveying in the Thar Desert with Mehboob, I decided to visit my Bishnoi friend in the Bap area near Phalodi, Jodhpur. Seated in his circular hut, my friend was enjoying a traditional *bhang-afeem* (cannabis and opium) session with his friends. After the customary salutations, I sat beside my host, while Mehboob stood at the entrance. Everyone knew about the *Godawan*, so I got good information on this bird from the group. Soon it was my turn to partake of the pot of *bhang-afeem*. In order to not embarrass my host, I took two *golis* of bhang (they tasted peppery, chewy) of bhang and kept them between my lower lip and teeth to avoid swallowing them. Mehboob slipped away, pretending to repair the vehicle engine. I sat for about an hour, discussing bustard conservation with my Bishnoi friends. The *golis* started to take effect, even though I had not swallowed them. Soon I was talking like Marlon Brando in the famous movie *Godfather*! (To those who are too young to have watched this, Brando stuffed Kleenex tissue in his cheeks to create the Godfather's distinctively guttural voice.) Mehboob rescued me before more *bhang* was offered!

In 2000, Rishad Naoroji (a Godrej scion interested in birds and conservation) asked me if Mehboob could work for him. The offer of both salary and accommodation was so tempting to Mehboob that, leaving everything aside, I let him accept it. Rishad was offering double the salary that Mehboob was getting from BNHS, plus accommodation in the Godrej Colony. It was the initial years of my directorship at BNHS, and I was busy learning administration, leaving very little time for fieldwork; it was time to let Mehboob go.

Once Rishad Naoroji took him as his personal assistant, Mehboob's life changed. Soon, he became Rishad's 'Man Friday', accompanying him on all his field trips, from Kerala to Ladakh, and from Rajasthan to Arunachal. In later years, when Rishad needed an assistant during his foreign trips, he took Mehboob to Europe and the US, and to Ecuador in South America and on a Galapagos cruise, among other places. After seeing endemic birds in Hawaii, Mehboob had telephoned me in excitement, not realizing that it was midnight in India! Mehboob later told me that Rishad treated him as an equal;

they stayed in the same hotels and ate the same food. Looking back, I am glad things worked out for him.

Some of the other remarkable field assistants I worked with include Adishisha in Rollapadu, Bhagwat Maske in Nannaj and Munni Lal in Dudhwa. These wonderful people not only helped me in the field but also shared with me their life struggles, family lives, financial problems, aspirations, agrarian and social crises.

As field scientists, our aim should be to make our field assistants' lives easy, mentor them, show them the right path and most importantly treat them with respect, dignity and equality. Whether I succeeded or not, I do not know. But I do know that I tried.

24

The Forest Job

During my career in wildlife conservation spanning nearly 50 years and counting, I have always had good relations with forest officers. I strongly believe that conservationists and conservation NGOs have to work with the respective forest and district officers to make an impact on the ground. We can give advice based on scientific research, but its implementation is done (or not done) by the authorities; and in case of Protected Areas, the forest officers are in charge. In the scientific world, there are 'the good, the bad and the ugly' scientists; similar is the case with the forest departments, but most officers fall in the first category. We also have to understand the constraints – political, social, administrative and financial – under which a forest officer has to work. Whenever I see that a certain conservation action has not been taken by a forest officer, I put myself in his/her place and judge whether I would have done it differently or better.

In the course of my story so far, you have met with a few of them, such as P.M. Lad and Pushp Kumar, though over the past five decades, I have interacted with hundreds of good forest officers. One of the earliest of the fraternity to leave a lasting impression on me was V.B. Singh, who was the CWLW of UP for many terms during the 1960s and 1980s. While studying in Aligarh, I would regularly write to him on conservation issues. Every visit to Lucknow, where my father was posted as chairman of the Sunni Wakf Board after retirement, would take me to Qaiserbagh and Nakkas bird markets, notorious for the sale of wild birds. In my broken English, I would write a flurry of letters to the CWLW, requesting him to raid those illegal markets. I would follow this up by meeting V.B. Singh at his 17, Rana Pratap Marg office. He would listen carefully and instruct his officers to take action against the illegal trade of wild birds.

In the 1970s, Kukrail forest was about 6 km from Lucknow, but rarely visited by anyone. I used to go there on my bicycle, with two or three friends, as going alone was risky due to looters and anti-social elements. In 1976, I wrote an article in *The Pioneer* newspaper, 'Bird Watching in Kukrail'[53] and gave a copy to V.P. Singh. I still remember how happy he was, and how he patted me on my back for knowing so many birds.

When I read that the Forest Minister of UP had announced a 'tiger safari' in Kukrail, I wrote an article, 'Why a Tiger Safari will not Work like a Lion Safari?', which was published in *The Pioneer*. My argument was simple: The tiger is a solitary territorial animal, and unlike the lion, prefers to dwell in cover and sleeps for 16–18 hours a day. In a safari park (where visitors sit inside a mobile cage [bus] and animals are free to roam in a large enclosure), visitors are interested in seeing animals such as the zebra, deer, antelopes, giraffe and lions, which live in big groups and are diurnal. Lion safaris are successful because lions live in the open in large prides, and thus can be seen easily – sleeping, sitting, yawning or doing nothing. In the case of the tiger or leopard this is not possible, as they are solitary and generally remain hidden most of the time. I wrote all of this in my article and sent the cutting to V.B. Singh. Since he too was against the concept of a tiger safari at Kukrail, he used my article to convince the Honourable Minister to drop the idea. Unfortunately, this absurd idea of tiger safari keeps coming up again and again in different states, even after 45 years!

Some of the best forest officers are the ones who offer unqualified support for research work; Dr Rupak De was one such field director. I had first heard of him through my student Salim Javed, who was working in Dudhwa for the Grassland Ecology Project from 1992 to 1995. He told me that the field director was extremely supportive of research; this was not surprising, as Dr De himself was a life science student, an MSc in Botany from Delhi University and had completed his PhD on angiosperms from the Vindhyan region of UP. His scientific background is reflected in all his official documents. His 'Management Plan of the Dudhwa Tiger Reserve' was so well written that the GoI circulated it to all states as a model for the preparation of management plans of protected areas all over the country.

I met him when he was the chief conservator of forests, but our friendship flourished when he took over as the CWLW of UP in 2012. Being a researcher with 15 scientific papers and numerous official reports to his credit, he was empathetic towards research work even back then. Dr De was also a strong pillar of my work on the Bengal Florican, and later the Sarus Crane. Affable, friendly and cooperative, I always found him in a cheerful mood in his large office on 17, Rana Pratap Marg, Lucknow. Fond of good tea, like me, he would never let me go without having tea and biscuits. The permissions for carrying out research in a protected area would be issued while having tea: the fastest permissions I have ever received from any CWLW!

My association with Dr Ram Dhirendra Rao Jakati was via Dr Vibhu Prakash. In 1999 or 2000, forest officers in Haryana had rescued a Cinereous Vulture (*Aegypius monachus*), and wrote to BNHS asking how they should treat it. As Vibhu Prakash was our raptor and vulture expert, he was sent to meet Dr Jakati, then the state's CWLW. During their conversation, Vibhu told him about BNHS–RSPB's plan to establish vulture conservation breeding centres in India. Dr Jakati promptly offered to start a centre in his state, and as they say, the rest is history. Later, other states came forward, but Haryana was the first to respond to BNHS' request. In the beginning, a Vulture Care Facility was developed on the land selected by Dr Jakati, and later it was expanded into the Vulture Conservation Breeding Centre on the same land at Pinjore.

Dr Jakati exhibited remarkable tenacity and efficiency in securing governmental backing for vulture conservation in the early 2000s. He personally crafted letters for clearances from the government, and engaged with colleagues and acquaintances to facilitate BNHS' nationwide vulture conservation efforts. Foreseeing potential issues in collaboration between a government department and an NGO, he facilitated a long-term Memorandum of Understanding between the Government of Haryana and BNHS for the Vulture Conservation Breeding Programme. He also played a pivotal role in establishing a Governing Council, led by the secretary, Department of Forests, Government of Haryana, which provided crucial support for the vulture centre's operations. Additionally, he chaired the Fund Raising, Advocacy and Communication Committee of SAVE, an international consortium of conservation organizations focused on vultures. Dr Jakati's unwavering support was instrumental in BNHS Vulture Conservation Programme's success – his 'baby' (the facility in Pinjore) has now grown into an internationally famous vulture breeding centre.

Dr Jakati succumbed to illness in 2022. In his passing, BNHS, vulture conservationists worldwide and I lost a true friend. I think the only way we can pay our respects to him is by ensuring that vultures repopulate the skies over India.

The name Dr Satish Kumar Sharma evokes double the joy in me, as two of my close associates share the name. One of them was my first PhD student, who is now professor and chairman in the Department of Wildlife Sciences, AMU. The second Satish is perhaps the most knowledgeable forest officer in Rajasthan. Now retired (officially) and settled in Udaipur, he represents the aphorism 'busy as a bee'. There is no 'unengaged' moment in Satish's life – for there is nothing that does not excite him, starting from fungi to baya to tribal culture to Marwari history. His knowledge of plants and their uses can fill many volumes of encyclopaedias, and his knowledge of Marwari history, architecture, legends, superstitions, erstwhile royal politics and literature can fill five or six PhD dissertations. On top of this, Satish is a mesmerizing speaker, with spellbinding stories to share with different age groups, from children right up to retired university professors.

I met him during the Dr Sálim Ali Centenary Seminar in 1996 in Mumbai, where he launched his book *Ornithobotany*. Satish told me that after the book release, a local Conservator of Forests (CF) from the Maharashtra cadre congratulated him. When the CF came to know that Satish was only a ranger, not an IFS officer, he was surprised and invited Satish home for dinner. For the occasion, the CF also invited the then Deputy Conservator of Forests, then Assistant Conservator of Forests and range forest officers from Maharashtra, so that they could be inspired by Satish. He recalls the CF telling his officers that they should 'either feel shame or have some inspiration [looking at Satish]'.

Starting as a range forest officer in 1980, Satish retired as assistant conservator of forests in 2016, giving 36 years of service to the Department, which is known for officialdom and lassitude rather than academic excellence. Besides being an active forest officer, and designing half a dozen new techniques of forestry, Satish took time to write 15 books, 350 scientific articles in Hindi and 292 articles in English. He has won 23 awards/honours, including the Indira Priyadarshini Vriksha Mitra Puruskar awarded by the GoI in 1999.

Some other officers showed dynamism in other distinct ways. One such individual was Rahul Bhatnagar, who, as chief conservator of forests (wildlife), Udaipur, started organizing annual bird fairs, which resulted in wide publicity for conservation, and also for the protection of many wetlands of the district. Most bird fairs in other states are just for entertainment, but Rahul saw to it that the Udaipur fairs resulted in better protection of natural habitats. As a result, wetlands that were earlier neglected or not well known got protection from the government, along with public support; in fact there is now a demand to declare one wetland, Menar, as a Ramsar site.

Attending the annual Udaipur bird fair rejuvenates me – Rahul has created great energy in the public for the cause of birds. How Rahul does all of this even after retirement and with limited resources is a mystery to me. Perhaps, his sincerity and a single-minded focus on the mission motivate people to help him. We need more officers like Rahul in India.

Until now I have written about senior forest officers with whom I worked and who impressed me with their dedication towards the cause of conservation. Many are retired and unfortunately a few are no more with us. Moving on to the younger crop of forest officers, whom I have known personally and often worked with, some names come to mind immediately.

One of those is Dr Mohan Ram, a bright young forest officer (IFS) of the 2012 batch, and presently posted as the DFO in Gir Forest (Wildlife division, Sasan Gir). After doing his MSc and PhD from the renowned Indian Agricultural Research Institute, New Delhi, he joined the forest service. In October 2019, I had the opportunity to go to Gir as a resource person for a Nature Guide Training, where I met Mohan Ram for the first time. I was impressed by his scientific understanding of the management of protected areas. When the officials at Gir informed me that they wanted to deploy satellite tracking devices on Lesser Florican, vultures and eagles, I advised them to engage Ali Hussain and his team. The catching operation was highly successful – Ali Hussain has not faced defeat till now – and 23 vultures and eagles were caught. From the results of the satellite tracking of birds, Mohan Ram and his team (which included representatives of TCF) have written seven papers on the movement of Lesser Florican, vultures, raptors and cranes, and many more are in the pipeline.

Unlike most forest officers, Mohan Ram is a good correspondent and quick to reply. Looking at him I feel that forest officers, with so much power and resources in their hands, can conduct excellent research and also facilitate research by others, much like the staff of USFWS. American journals are full of research papers by forest officers working in the USFWS and other government organizations. What stops us?

Research is not just the publication of papers in high-impact, peer-reviewed journals. Research teaches us how to understand the problem by analysing the data, reading earlier work on the topic, asking questions, discussing with peers and experts, and then coming to a conclusion. What we lack in India is research-based conservation management. In the state of Gujarat, I have seen the destruction of the wonderful Khijadiya Bird Sanctuary by forest officers, whose understanding of wetland ecology was below par. Fortunately, we have young forest officers like Mohan Ram, in the same state, who are strong in science and administration, and hopefully will not repeat the same mistakes that their seniors had committed in the Khijadiya wetland of Jamnagar, or the Naliya grassland of Kutch.

In 2000, the Union Public Service Commission, New Delhi, requested me to be on their interview panel to interview IFS candidates who had passed the written tests. I think there were eight members on the panel, and among them I was the wildlife 'expert'. I found that candidates with wildlife or life science background did very well, and one of them was Sonali Ghosh. Even now, whenever I meet her, she tells me, 'Sir, you had selected me and given the maximum marks.' Of course, I can't take the credit for it, as she did well in the interview. I was delighted to hear that in 2023 Sonali became the first woman field director of the Kaziranga National Park.

In her 23-year career, she has held many key positions, including assistant field director of Manas Tiger Reserve. She has travelled extensively in Northeast India, and has also interacted with the local people. Nearly 15 years ago, we travelled together to the Ultapani forests that were under tremendous pressure due to encroachment. The Ultapani Reserve Forest, famous for its butterfly diversity, is situated in the Holtugaon Forest Division of Manas Biosphere Reserve, Kokrajhar district, Assam. The name means 'reverse water', as the forest river flows west to east, unlike other rivers that flow from east to west in that region. As DFO of Kokrajhar district, with the additional charge of wildlife from 2007 to 2009, Sonali undertook a comprehensive Golden Langur

(*Trachypithecus geei*) survey across its entire distribution in the country. She also took up the rescue and rehabilitation of two Clouded Leopard (*Neofelis nebulosa*) cubs back into the wild, which was the first-ever such recorded case for the species. This was done with the help of NGOs in the Ripu Reserve Forest, after more than a year of housing and training them.

Scientific data from such initiatives helped strengthen the evidence for declaring the rich forests of Kokrajhar district as Raimona National Park in June 2021. Raimona includes both Ripu Reserve Forest and Ultapani forest. The area, which is part of the larger Manas Tiger Reserve landscape, is known for its unique biodiversity and provides a crucial habitat for many species, including elephants, tigers and a range of flora and fauna. The inclusion of Ultapani and Ripu forests in Raimona National Park helps in the conservation of these critical habitats and the species that depend on them.

I had the privilege of working on books written by three forest officers. One of them is Dr Dhananjai Mohan, who first wrote to me in 1997, when he was posted at the Katerniaghat Wildlife Sanctuary (UP), a place I visited many times during the Florican Project in the 1980s and 1990s. We connected due to our mutual interest in birds; I was particularly happy at the thought that there was now a young forest officer with whom I could discuss bird and grassland conservation in UP. Dhananjai is a B.Tech. in electrical engineering from the Indian Institute of Technology, Kanpur, but his interest in nature prompted him to join the IFS in 1988. In the initial part of his service, he served in the high Himalayan protected areas in undivided UP, and later worked in the Forestry Research Division and in the Katarniaghat Wildlife Sanctuary in the terai region of the state. He made a detailed checklist of birds of Katerniaghat, which impressed me very much.

After the split of UP in 2000, which led to the formation of Uttarakhand, Dhananjai shifted his base to the new state, where he continued the avifaunal work, primarily in the Himalayas. Many a time, we planned to survey areas where the Mountain Quail (now feared extinct) could be seen, but the plan could not materialize. We also exchanged notes on Himalayan birds; actually, it was mainly a one-way traffic, as my knowledge of Himalayan birds was meagre and I learnt a lot about Himalayan avifauna from him. Dhananjai has nearly three dozen publications, primarily on the biological and management aspects of biodiversity conservation, of which half pertains to birds. After a seven-year stint at the WII, where he was spearheading avifaunal research, he went back to the Forest Department, becoming the chief conservator of forests (wildlife) of Uttarakhand.

In 2011, we decided to co-author the book *Threatened Birds of Uttarakhand*, which came out in 2013. Interestingly, during the writing of this book we never met, though hundreds of emails kept us connected. I was very happy when he joined as director of WII, but his stint was short-lived, as he was called back by the Uttarakhand Government.

It was with another forest officer, Dr Manoj V. Nair, that I co-authored *Threatened Birds of Odisha* in 2012. An avid naturalist since childhood, Manoj has over 25 years of experience in watching birds. He obtained a master's degree in wildlife science from the WII and joined the IFS in 2001. Author of six books and about 50 scientific publications since

1990, he was awarded the Kailash Sankhala National Wildlife Award (given by the MoEF, GoI) for the year 2012.

Another book was written with Dr Gobind Sagar Bhardwaj, a forest officer who made national news in 1997, when he saved villagers near Ranthambore from a tiger that had strayed outside the forest. In the encounter, Gobind was badly mauled and sustained near-fatal injuries. He would later write to me: '[The] tiger attack incident in [...] 1997 and my subsequent PG diploma in Wildlife Management made me interested in the field of wildlife conservation. My tenure as [the head] of Ranthambhore Tiger Reserve was the real turning point [in] my career. Later, it was my PhD that took me towards bird conservation and birding as a hobby.'

I met Gobind during my visits to the WII in the the early 2000s, and was impressed by his knowledge of plants, birds, ecosystems, protected areas and general ecology. Conversations with him were never boring, because of his friendly nature and good humour. He is from Himachal but speaks in a Punjabi accent, which makes him even more endearing, fond as I am of the Punjabi language, particularly high-spirited Punjabi songs. Gobind's writing skills and photography also stood out.

During his tenure as scientist in the WII (2008–12), Gobind did extensive surveys of the Lesser Florican in its distributional range, in addition to the avian faunal surveys in the trans-Himalayan region. Later, he served as the field director of the Sariska Tiger Reserve. We became close friends when he was the chief conservator of forests (wildlife), Jodhpur, and was supervising various in situ conservation efforts for the critically endangered GIB. In 2014, when I took a team of BNHS researchers to show the areas where I had seen bustards from the 1980s onwards, we met him in Jodhpur many times, and he also accompanied us to the Desert National Park. While sitting in the hut outside Desert National Park's Sudasari enclosure, we decided to jointly write a book on that neglected park. Our book, *Desert National Park: A Jewel in the Vibrant Thar*, came out in 2020; it was funded by the Hem Chand Mahindra Foundation and sponsored by TCF. The discussions on the manuscript, via email and telephone calls, illuminated the dark days of Covid. Another joint book on the Ladakh landscape is what we are working on right now.

Books, the love of a landscape or species – these are the bonds that bind me to the brotherhood of forest officers in India. Sitting with the forest officers in a remote forest guest house, exchanging wildlife knowledge over hot cups of tea or coffee, understanding their difficulties, the demands of local communities, and the political and societal pressures on them, has made me realize how difficult their job is. Compared to their work pressure, particularly if the officer is honest and dedicated, wildlife research is a cake walk. And much more rewarding!

25

Colleagues in Conservation

During my wildlife career, I have met some of the most amazing people connected with the conservation world. Some started as my students but are now valued colleagues, some worked with BNHS during my two tenures and a few were closely associated with BNHS or conservation for a long time. Each one had an influence on my life and on the field in which they were working.

Let me start with a duo who began as my students. The Department of Wildlife Sciences of AMU has produced some of the finest wildlife biologists of India, and Qamar Qureshi and Parikshit Goutam are two shining examples. I met them in 1984, when I went there to give a lecture to BSc students. Even at that time, they had stood out in a class of nearly 40 students. In 1988, armed with newly acquired MSc degrees, they accompanied me on a five-week survey of the terai in UP, along with Bihar, West Bengal and Assam. As I wrote earlier, I did not have good knowledge of statistics – the subject in which Qamar was well-versed. During the survey in 1988, I learnt a few basic principles of statistics from him, followed by a full-fledged course in the summer of 1989 when Qamar came to Dudhwa to study Swamp Deer (*Rucervus duvaucelii*) for his doctorate. Qamar was soon selected by the WII, where he now works as a senior scientist, while Parikshit, after working for a few years in WII, joined WWF-India where he worked for more than 20 years. He is now in China, working as a senior consultant.

There is a generational gap between me and Aasheesh Pittie, but what binds us is the love for books and birds. Aasheesh is a bibliophile and perhaps has the finest bird book collection in India. I first met him in 1981, when he was still an unmarried young man, full of energy and unquenchable thirst for knowledge. Pushp Kumar (CWLW of Andhra Pradesh) introduced him and Siraj Taher to me in his office; Aasheesh and Siraj had established the BSAP, now called the Deccan Birders, and Pushp Kumar wanted me to meet them. We hit it off well, due to our common interests. In the 1980s and 1990s, during almost every visit to Hyderabad, Aasheesh and the late Siraj Taher would arrange for me to give a lecture to the members of the BSAP. Unfortunately, Siraj left us in 2010, but the tradition of lectures continues; the latest one in 2023 was on the 'History of Natural History in India'.

During one such visit to Hyderabad, I think in 1983, Pushp Kumar asked me to meet a young crocodile researcher named B.C. Choudhury. I had heard of him, as I knew about the Crocodile/Gharial Project started under Dr H.R. Bustard, a Food and Agriculture Organization (FAO) expert sent to India to initiate crocodile conservation in the early 1970s. I spent nearly two hours with B.C., a meeting that developed into a life-long friendship. In the 1990s and 2000s, we would meet frequently during the official meetings of MoEFCC, and for four years B.C. was on the Governing Board of BNHS while I was the director. Now he and his wife Nalini, who is like a younger sister to me, are settled in Dehradun. B.C. retired from the WII as a professor, and like a true professor, his knowledge is vast. We have been in the field together several times, in Tamil Nadu, Kerala, Gujarat and Rajasthan, mainly for wetland conservation – the neglected subject that binds us.

Another younger friend I often reach out to is Goutam Narayan; he is eight years younger than me, but whenever I need advice in a difficult situation, I call him. He was the most intelligent, rational, astute and sharp-minded colleague that I had in BNHS. Even after he left BNHS in 1991, we meet whenever I go to Assam, where he currently resides.

In the 1980s, we worked together on the Bengal Florican project in Manas, and in early 1995, he took charge of the Pygmy Hog Breeding Programme in Basista, near Guwahati, Assam. I was on its advisory board, and got more opportunities to interact with him. Under his supervision, and with the support of the Assam Forest Department and the late conservationist William Oliver, the Pygmy Hog project (which manages the captive breeding of the rarest wild pig in the world) has been very successful. The Programme has seen the successful establishment of populations of this small pig – about the size of a cat – in Manas, Orang, Bornadi and

Sonai Rupai grasslands, habitats that Goutam and I have surveyed many times.

Since I always give attention to neglected species, ecosystems and regions, I was pleasantly surprised when in 1994 I heard about the Mandar Nature Club, an organization established in Bhagalpur, Bihar, by a young greenhorn, Arvind Mishra in 1990. Later, in 1996, I heard about him again, when I was working in Aligarh as the chairman of the Centre of Wildlife. *The Pioneer* carried an interview with Arvind, and talked about how while the Department of Animal Husbandry and Fisheries, Government of Bihar, had prohibited fishing leases in two bird sanctuaries (Nagi and Nakti) in 1993, the leases were still being officially offered, through advertisements released by the local Irrigation Department. When I read the article, I made several recommendations to the then state environment minister, Capt. Jai Narain Prasad Nishad, and decision-makers in Bihar to stop fishing and instead make the two wetlands into bird sanctuaries. I also took a chance and sent copies of my letter to Arvind, asking for his comments and updates on the issue. He was surprised at this and quickly replied in his 'Hinglish'. That was the start of my friendship with Arvind, an extremely dedicated conservationist of Bihar, who is a true 'salt of the earth'.

When we established the IBCN in 1999–2000 and started working on the IBA project, Arvind was our first choice for Bihar's state coordinator, a role he played with great aplomb. In 2006, he excitedly contacted me to tell that he had found a colony of Greater Adjutants in Bhagalpur district. I was sceptical, and asked for photographs, which Arvind sent by email; those convinced me that he had indeed found a colony of this endangered species (it is included in IUCN's Red List). Arvind told me that the villagers knew about the bird, as it bred on tall trees and foraged in their crop fields, but the population was unknown to conservationists. Locals considered it a useful bird, as it ate the rats and snakes that emerge when the farmers flood the fields for rice cultivation.

In 2010, along with Arvind, I visited the *doab* area near Bhagalpur where these birds breed. *Doab* is the land between two large rivers that usually gets flooded during the monsoons. Arvind knew about the location of most of the nests, and so we spent the entire day going from one nest cluster to the other. I was impressed by his knowledge and the way he was received warmly by villagers. I told Arvind to keep records of nest sites, breeding success of each nest/colony and the total population of Greater Adjutants. I contacted Dr Hilloljyoti Sangha, my former PhD student, who had conducted research on this species, and Purnima Barman, who had started working on Greater Adjutants in Assam, to share the exciting news that a fairly large colony of this 'Vulnerable' (Red List category) species was thriving in Bihar.

During this visit, I saw farmers sowing or weeding crops, with Greater Adjutants standing close, to pounce on an escaping rat or snake. Thanks to the efforts of Arvind and the support of the local people, more than 300 Greater Adjutants are found in Bihar now. In the 2018–19 breeding season, they were able to raise 150–160 juveniles. Spread over a few hundred square kilometres, the *doab* area in Bhagalpur is the second largest breeding colony of this bird in the world. I mainly credit Arvind for the protection these birds have been able to secure, though both the Forest Department and the local communities have also played a huge role. To me, Arvind is the perfect example of what passion and commitment can do for a cause.

⁓

I first met Manoj Kulshreshtha in 1988, at the house of Bharat Singh (the *sarpanch* of Kundanpur), who lived in Kundanpur village of the then Kota (now Baran) district of Rajasthan. At that time, Manoj was documenting the bustard population and its habitat, with the support of an army infantry division led by General Commanding Officer Maj. Gen. Baljeet Singh, who is a life member of BNHS. Manoj, an MSc in Agriculture Sciences, was then working with the Rajasthan Agriculture University as a plant breeder. Our friendship of 35 years has flourished thanks to our mutual interest in conservation. Since 1994, Manoj has been my companion in all Thar Desert surveys.

Our companionship is symbiotic – he loves fieldwork, and I get the company of a very resourceful person, co-driver, friend and a researcher in the field. He speaks several Rajasthani dialects, knows local customs, etiquettes and cultures, and has friends and relatives in almost every town of the state! He also speaks with a sincere passion to local people, thus connecting with them easily. For environmental communication, knowing the local language, customs and cultural norms are extremely important. When we started the IBA programme and wanted to organize a workshop for collecting information on potential IBAs, Manoj was the first person who offered to organize the workshop in Rajasthan. He was the state coordinator of IBCN as long the programme survived.

In 2000, during a *padyatra* (or walkathon) from Bikaner to Jaisalmer for *Godawan* (bustard) conservation awareness, Manoj made most of the arrangements, and when we wanted to put up a bustard *moorti* (statue) in Sankhala village, Jaisalmer, Manoj got it sculpted at Jaipur, sitting with the sculptor for many days to ensure that the size and posture of the bird were correct. Manoj also helped me organize a meeting on the 'Future of Desert National Park' at Jaipur.

Besides our interest in nature conservation, we share an interest in old Hindustani songs, ghazals, mutton biryani, movies, as well as in our political leanings. Manoj is so fond of ghazals that he is learning Urdu. I sincerely hope that he does not start sending me emails in Urdu, a language I can barely read!

While working on my book *Threatened Birds of India*, sometime in 2010 or 2011, I came across a picture of the Black-breasted Parrotbill (*Paradoxornis flavirostris*), a vulnerable species, taken by a man called Dhritiman Mukherjee. Wanting to use the image in my book, I wrote an email to him. There was no reply for some months so I wrote to him again, offering money for the one-time use of his picture. No reply again! I was getting excellent bird images from many professional photographers, mostly for free and also quickly, so I thought Dhritiman a rude fellow who did not want to share his images of birds. One day, I asked my colleague Isaac Kehimkar, who had returned from West Bengal and Assam, whether he knew Dhritiman. 'He is a good photographer but arrogant, as he does not reply,' I added. Isaac promptly sent an email to Dhritiman about my grievance. It then turned out that I was emailing Dhritiman using an address that was no longer in use. As soon as I reached out via the correct email, prompt came the reply, offering images of many threatened species. Books talk about 'love at first sight' but for me it was love at Dhritiman's first email! He was kind, loving, cooperative and friendly – the perfect friend! So different from the image of the 'rude, arrogant' man that I had built up.

After a few months, Dhritiman came to BNHS. Initially, he was afraid to meet me (given the background of our 'encounter'!) but when I came to know that he was sitting with Dr Raju Kasambe discussing IBCN and IBA activities, I went to meet him and brought him back to my chamber. To break the ice, I took him to my favourite restaurant 'Chetna', where we ordered two Rajasthani *thali*s. I soon realized that he was not eating, but playing with the tasty sabzi, dal and bajra rotis. Eventually, he was able to 'finish' his meal, though with great difficulty, and half the food remained on the plate. I shared my *gulab jamun*, which he gladly ate, proving his Bengali fondness for sweets. Looking at the hefty, 6 feet tall young man, I guessed that he was missing his fish curry and rice (Chetna is a strictly vegetarian restaurant). Much later, I found out that Dhritiman is a strict carnivore, who hates vegetables and fruits.

Despite the fact that our first joint meal was a disaster (for him, not me) – plus I was quite cut up with him for wasting my Rs 400 on the *thali* – our friendship cemented as more meals progressed. I found him a sincere, dedicated and socially conscious person. He asked me to take him to Andaman's

Narcondam Island, where I had gone a year earlier on a short one-day visit. Our trip, as detailed in an earlier chapter, fructified, and further underscored the kind of visual nature historian that Dhritiman is.

Since that field trip to Narcondam in March 2013, we have gone on many trips together, from Rajasthan and Gujarat to Ladakh and Assam. Most of my articles and books have Dhritiman's images, and in 2015, we co-wrote *Magical Biodiversity of India*, which was an instant success and got sold out soon.

My first meeting with Neeraj Srivastava did not start on a good note. In 2006, BNHS–RSPB had organized an IBCN State Coordinators' Meet outside the Corbett National Park. Aqeel Farooqui, being a UP state coordinator, was invited. He called me, requesting to bring his colleague and friend, Neeraj Srivastava, along. I had severe reservations about this, as we had limited funds and the venue was quite expensive and also because it would set a wrong precedent. Aqeel pleaded his case, stating that his friend was a very committed conservationist and that he would not be a burden on the organizers, logistically or financially! Owing to his persistence, I reluctantly agreed.

However, one meeting with Neeraj and I knew we had found the person we were looking for to champion conservation in UP! Aqeel and Neeraj were colleagues in the UP State Road Transport Corporation, and both were interested in wildlife. Both were also solid communicators. I co-wrote the *Threatened Birds of Uttar Pradesh* with Neeraj, and we are planning another book together. Aqeel has written the eminently readable *Ankush Days: The Story of Ishtiaq, Mahout of Corbett Tiger Reserve*, for which I wrote the 'Foreword'.

Besides the students, colleagues and friends that I have described above, there are many more people who have been an important part of my life. Raghunandan Chundawat, Joanna van Gruisen, Bishwaroop Raha, Ashok Kumar, Belinda Wright, Intesar Suhail, Khursheed Ahmad, Delip Kumar Das, Bivash Pandav, Anish Andheria, Vivek Menon, Sunjoy Monga, officials in the MoEFCC... the list is long and my recollections on all of them would fill a book in itself!

26

The 'Babu' as Conservationist

I secured 94 per cent marks in mathematics in high school, based on which my father suggested that I become an engineer. A few years later, my eldest brother, Khalid, wanted me to appear for the IAS exams. Fortunately, I took my own decision and became a wildlife biologist and lived happily thereafter, as the cliché goes.

Since childhood, though, I have seen the power of a district collector, who has *rutba* (Urdu for status, prestige, power that someone has, in the eyes of other people). During my childhood, the district collectors were my father's friends, 'uncles' who lived in large bungalows, with a flock of liveried servants. I remember my father taking us to see the National Academy of Administration in Charleville Estate, Mussoorie, where IAS officers were trained. Now it is known as the Lal Bahadur Shastri National Academy of Administration (LBSNAA). After nearly 55 years, I visited LBSNAA in April 2019. It was like stepping back into the past. I must add here that LBSNAA has one of the cleanest campuses and buildings in India. Its library will shame any university's library. It has 187,485 documents, including 160,734 bound volumes of journals; 8,679 audio cassettes; 2,109 video cassettes; 1,708 CDs/DVDs; and 7,996 lecture recordings, documentaries, movies in different languages, and digitized rare and old books. No wonder many IAS officers are so well read! But, unfortunately, not all.

At the state level, wildlife conservation is considered the domain of the Forest Department, and rightly so, as they are the custodians of our protected areas; they enforce the Indian Wildlife Protection Act (though police have been given power to control poaching and other violations) and to some extent the Forest Conservation Act. Nowadays, their numbers are supplemented by a new breed of official conservationists, whom I call the 'bureaucrat-conservationists' – members of the IAS – who are contributing greatly to the conservation movement in India. Some of them have played a key role in the establishment of a sanctuary or taken steps towards the protection of a species.

Sanjay Kumar is an IAS officer belonging to the UP cadre. How do I define him? Conservationist first, bureaucrat second or vice versa? Or as a young, able administrator, birdwatcher, writer, photographer and a great humanist? I have known Sanjay for the last 10 years. Wherever he was posted, he made a difference in the wildlife scene of his district. For example, when he was posted as district magistrate of Sitapur, he worked for wetland conservation and brought out a very informative book, *Conservation of Potential Wetlands in District Sitapur*. He developed a model of wetland conservation by involving the local community. Around 30 wetlands spread over 700 ha were restored, as a result of which the migratory birds returned to the area after a long time. He also mapped the district's avian diversity, and conducted surveys of the Sharda and Ghagra rivers along with experts, documenting the biodiversity around the rivers for the first time.

Sanjay worked hard for the restoration of natural habitats, particularly wetlands and rivers. In Moradabad, for example, he tried to clean the rivers, and in Allahabad he got the Chand Khamaria Blackbuck Conservation Reserve near Meja notified. It was the first such conservation reserve of the state to be notified by the Forest Department, and it happened due to the initiative of Sanjay! The Reserve's 300-odd acres of savannah landscape is home to over 350 Blackbuck.

Sanjay strongly believes in documenting everything, even the birds that are found in district magistrates' sprawling residences. He has written 10 books or booklets – including *Birds In and Around Mussoorie* (2013) and *Bird Diversity of Chandrashekhar Azad Park, Allahabad* (2017) – and is working on the eleventh one. His latest book, *Birds of Lucknow*, with Neeraj Srivastava is an excellent example of the need to bring out more such city/district-level books that not only have bird description, status and pretty pictures, but also information about the various habitats where a particular species is likely to occur, such as rivers and streams, wetlands, large gardens, lawns around monuments, temples and graveyards, old trees that attract birds, etc. Not many bird books have such rich information.

Perhaps his greatest contribution is that he motivates his fellow district officers, who are otherwise surrounded by files and red tape, and do not have the time to look into environmental issues or to work towards nature conservation. I also cannot forget the way he helped me when I had Covid (more on that later).

Another officer whom I much admire is Krishna Kumar Dwivedi of the Assam cadre (what a wonderful state to work in, with equally wonderful people). I first met him in Tinsukia in 2005, where he had already made a mark by working towards the protection of the much-neglected Dibru–Saikhowa National Park. He first brought everyone together – forest officers, local NGOs, local conservationists, intellectuals, photographers and school children – to highlight the plight of this lesser-known Park. Then, through his frequent visits to the Park, he was able to galvanize its somnolent administrative machinery. With the help of ace photographer Dhritiman Mukherjee, he wrote a coffee-table book on Dibru–Saikhowa, with a foreword by the President of India, Hon'ble Pratibha Patil. Even 18 years after his posting out of Tinsukia and Dibrugarh districts, people still remember him for his affable nature and his immense contribution in highlighting the Dibru–Saikhowa National Park. A prolific writer, he has published two books on the mighty Brahmaputra River, and recently edited a coffee-table book on the heritage of Assam tea.

Like Krishna Kumar Dwivedi, Vikram Singh, who retired from the Rajasthan cadre, was excellent at public outreach, and had a deep love for nature and commitment to nature conservation. I connected with him 10 years ago when he invited me to the first Bird Fair at Dungarpur, a beautiful dry district of south-western Rajasthan, full of ancient man-made lakes and ponds. Vikram Singh was the first bureaucrat to recognize the ornithological potential of the Dungarpur lakes, and he encouraged the local people to protect the birds (some were already doing so due to religious reasons). His first Bird Fair in 2013 became a role model for neighbouring Udaipur, where it became an annual feature since 2014, creating mass awareness. Now, Dungarpur and Udaipur districts compete with each other over which one has better-protected wetlands and which wetland attracts more birds. Wherever Vikram Singh was posted – Dungarpur, Banswada, Rajsamand, Chhitorgarh, Bhilwada – he had left behind a legacy of bird fairs.

His strategy involved working with district-level forest officers to consolidate protected areas, involving local communities and students, and arranging funds for conservation measures from existing schemes. He also worked closely with the local police to prevent poaching. In short, Vikram Singh is a one-man army and does the kind of work that a well-run conservation organization does.

Dr Anwaruddin Choudhury has a PhD, a DSc, and is also the author of 27 books/booklets, 45 technical/survey reports, and nearly 700 articles and scientific papers in national and international journals. You may think we are discussing the curriculum vitae of an old university professor, but no, Anwaruddin was a bureaucrat from the State Administrative Services in Assam, till he took VRS a few years ago, at the age of 58, to immerse himself completely in reading, writing and fieldwork. Not many people know that he is an artist and cartographer, and that all his books are illustrated by him. His first PhD was in geography and the second in zoology.

People who know this legendary conservationist from Northeast India – and there are legions of Anwar's admirers – actually joke that when he gets time from nature study, he does his official work! The fact was that despite holding several senior positions in the state government, he managed to go to the field, read, and write books and research papers. One should learn time management from Anwaruddin!

Anwaruddin's contribution to the development of protected areas will need many pages to document, for he has worked for the conservation of several sanctuaries and parks, such as the Nambor-Doigrung Wildlife Sanctuary (Golaghat district), Marat Longri Wildlife Sanctuary, North Karbi Anglong Wildlife Sanctuary, East Karbi Anglong Wildlife Sanctuary, Nambor Wildlife Sanctuary (all in Karbi Anglong district), Hollongapar Gibbon Sanctuary (Jorhat district), Bherjan-Borajan-Podumoni Wildlife Sanctuary (Tinsukia district), Dihing-Patkai National Park (Tinsukia and Dibrugarh districts), Barail Wildlife Sanctuary (Cachar district), Amchang Wildlife Sanctuary (Kamrup Metro district), Bordoibam-Bilmukh Bird Sanctuary (Lakhimpur and Dhemaji districts) and Pani-Dihing Bird Sanctuary (Sivasagar district).

He was also the greatest contributor to my book *Important Bird Areas of India* (2004) and its revised version *Important Bird and Biodiversity Areas of India* (2016). Except for Sikkim, the text for all six Northeastern states was written and edited by him. I also had the privilege of authoring a book with

Anwar, *Threatened Birds of Assam*, which went for reprint within six months; such is the popularity of his writings.

We have been friends for the last 35 years and communicate frequently on natural history issues. We have also done joint surveys in areas such as Manas, Kaziranga, Dibru-Saikhowa and Tinsukia. Ever smiling and soft-spoken, Anwar is truly an excellent example for the younger generation.

The work of other such bureaucrat-conservationists is not as well known. Sometime in the early 1990s, I was introduced to Pravinsingh Pardeshi by Bharat Bhushan, my former colleague in the bustard project. Later, I came to know that he was posted as the collector of Solapur district, where the Nannaj area of the bustard sanctuary is located, and where I had worked in the 1980s with Ranjit Manakadan. Pravin worked for the protection of bustards and grasslands in Solapur, and helped the Forest Department identify and consolidate many grassland patches in the district. Pravin is well known for his excellent rehabilitation work after the Latur earthquake in September 1993, and his contributions as chief of the transition recovery unit of the United Nations; but his conservation work is less known.

These are but a few of the many hundreds of good officers in various departments – administrative services, police, Border Security Force, Indo-Tibetan Border Police, the three wings of the armed forces – contributing towards wildlife and nature conservation. Of course, they are not doing it for popularity, but a little recognition of their contribution would motivate others. Our wildlife needs support from everyone, especially those with the power to set aside spaces for wild animals to live in and protect them from poachers. And who better to do that than a top bureaucrat of a district or a state.

27

Moving on From BNHS

My great friend B.C. Choudhury jokes that the only event of our life that we can be sure of is the date of superannuation/retirement. Or, as others put it: there are three certainties of life – taxes, retirement date and death!

One year before my 30 July 2015 retirement, I started preparing for the transition by giving much more freedom to the senior staff, deputing them to represent BNHS in national and international meetings, and asking institutions and people to correspond directly with them, thus, slowly distancing myself from the day-to-day functioning of BNHS. Raju Kasambe, for example, started representing BNHS in BirdLife Asia meetings, and Shubhalaxmi began taking independent decisions regarding CECs of Mumbai and Delhi. Dr Balachandran, meanwhile, was independently looking after BNHS' bird-ringing activities, and also training younger staff. For administrative purposes, Deepak Apte was made the chief operating officer. Meanwhile, my personal assistant, Sachin Kulkarni was instructed to catalogue and archive important papers of my tenure. Also, in order to continue programmes on the GIB in the Thar Desert, where its numbers were declining fast, in February 2014, I took Sujit Narwade and four more staff to introduce them to my contacts, and show them the areas where I had seen bustards.

I withdrew as principal investigator of projects, which were funded in 2014–15, and senior scientists were asked to take charge. However, it was also not possible to withdraw completely from some projects. I continued as principal investigator, for example, in the Bengal Florican Project. However, my aim was to see that the work continued without any vacuum after my retirement. I can say that I was partially successful. The BNHS board also asked me to continue as a senior scientific advisor to the president and director of BNHS for two years, until the transition was complete.

When the D-day came on 30 July 2015, the staff threw a big party, along with gifts and mementos. Some became emotional as they spoke, some cried, but overall, the occasion was full of bonhomie and fun; a highlight was a wonderful skit about my likes and dislikes. That day, I came to the office in the official vehicle with my favourite driver, Paras, and went back home in a taxi. Paras was in tears as he shut the taxi door. My official association with BNHS had ended after 35 years. But did it really end?

For me BNHS is like family, a part of me, my life. Post retirement, BNHS allowed me to shift to a small room with my books, personal files and computer, where I could work as a senior scientific advisor. I would go to the Hornbill House only once or twice a week, and work on my incomplete book, *Important Bird and Biodiversity Areas of India: Revised and Updated*. The most amazing part was the love, respect and affection of the BNHS staff, which did not diminish even after my retirement, and as I write this in July 2024, I see no change now either. Whenever I go to BNHS, it seems as though nothing much has changed. The new staff appointed in the last nine years do not know me well, but the old staff have the same love towards me. The fondness, respect and warmth that we gave to J.C. Daniel after his retirement in 1991 till his death in 2011, has been transferred to me. This is the strength of the BNHS family.

As a scientific advisor to BNHS, I decided to stay back in Mumbai, but finding a rented apartment was a nightmare. More than any financial constraint, the prejudice towards the 'other community' (in other words, my religious background) became the reason for rejection. After meeting or contacting many apartment owners, finally M.G. Mathews, BNHS's administrative officer, helped me find a good apartment in the Daffodil Housing Society in Borivali. It had a mix of people from various communities – Marathi, Gujarati, Konkani, Goan – and different faiths – Christian, Hindu and Muslim. My rented apartment faced a vast swath of mangroves in the distance. Amjad, my factotum or assistant for the previous 18 years, stayed with me, doing double duty, as he was employed by the central marketing department of BNHS.

I decided to move all my books and papers to my new apartment, and would spend most of my time reading and writing. Once or twice a week, when I went to BNHS, I would commute by the local train. As my family lived a nomadic life, with my father being transferred every two to four years, and because I love travelling and fieldwork, adjusting to the new life was easy. Some of my colleagues were surprised to learn that after travelling in a chauffeur-driven SUV for 18 years, I now travelled in local trains. But for me, it was simple: the vehicle and the driver were given to the director, not to me, so when I relinquished the post, I relinquished all the perks. My leftist propensities also helped me adjust quickly – mentally I was always a part of the hoi polloi, not any privileged class. I started enjoying taking an early morning bus to the Borivali station, a brisk walk to the railway platform followed, with furtive glances at the minute hand of the watch. The morning's tempo picked up with a rush towards the incoming train, and finally one had the satisfaction of getting a seat! Nearly 7.5 million Mumbaikars do this every working day, so why should I be any different from them? I love being a normal Indian citizen, rather than a privileged one.

Even before my retirement, Dr Deepak Apte had been selected as the director. I was happy that the transition was smooth and that I was replaced by a person who knew the workings of BNHS. There were some differences of opinions between us while I was the director but I made sure that they did not come in our way, and that the overall functioning of the Society was not impacted.

Even though I was a paid scientific advisor for two years, I was never asked for any advice, even on the subjects on which I had deep, practical knowledge! However, I never interfered in the administration and did not encourage the staff to say anything against the new administration. I would always say, 'Things change, different people have different ways of working, so accept the change.' Gradually, I further withdrew myself by working from home, giving space to the new director. By 2018, I decided to shift to Lucknow from my rented apartment in Borivali. It was not an easy decision, as many of my friends were in Mumbai, but living there was becoming too expensive. Moreover, I had already bought an apartment in Lucknow in 2008; so I thought it better to move there and restart my life. The presence of close relatives and friends, such as Neeraj Srivastava and Sanjay Kumar, in Lucknow also influenced my decision.

Like everyone else, my life too is not without its share of unfulfilled dreams, incomplete missions and personal regrets. My autobiography will not be complete unless I write about my regrets, not in any order of importance, but nonetheless worth mentioning.

One of my greatest regrets is not learning the beautiful language of Marathi. I spent almost 40 years in Mumbai, Maharashtra, but never learnt to speak Marathi (though I can understand it). In the first phase of my BNHS life (1980 to 1991), I was in and out of Mumbai but from 1997 to 2018, I was mostly in Mumbai; so I have no excuses other than to blame my schedules, which always kept me busy. I know this is a puerile excuse but this is what I can write to acquit myself of my mistake. My experience in Mumbai, I suspect, lacked a little colour, as I did not learn Marathi. Though Hindi is like Esperanto – the universal language proposed in 1887 by L.L. Zamenhof and his wife – in Mumbai, it is always good to communicate in the local language with the local people, particularly in the rural areas.

Another regret is about not understanding statistics as much as I would have liked to. When I did my MSc in Zoology in the early 1970s, biostatistics was not a part of the syllabus. Later in the late 1980s, I learnt basic statistics from Qamar Qureshi, while staying in Dudhwa where he was also working. While he and Dr Y. Jhala are two of the finest field biologists of India, with clear understanding of where to use statistics (overuse of statistics by confused biologists is now quite common), it is a challenge to grasp the various statistical formulae and programmes that are commonly used in good research papers these days.

One more regret is that I did not write many research papers on time. I can pacify myself, inanely, that I was too hard-pressed for time but writing a good paper, and on time, is the job of any good scientist and field biologist. In the conservation field, papers are not to satisfy academic requirement or one's ego, but to spread the message of conservation to decision-makers, so that they could take appropriate actions for saving species and habitats. I think it is the duty of conservation biologists to bring out papers and reports on time, and to disseminate them to concerned authorities. We fail if we do not do this. Additionally, our duty towards funding agencies requires us to broadcast our findings to the public and the government. Although I have written nearly 180 research papers, numerous short notes and more than 350 articles, I feel I could have done more. I think, just as there is never enough conservation activity, similarly, there are never enough publications.

Many doctors suggest that to keep the mind strong and alert, and to prevent dementia in old age, we need to keep our brain engaged by learning new techniques, new languages and by developing new hobbies. I still read and write every day, and maintain a daily diary and field notes whenever I am in the field, but perhaps I should learn Marathi and statistics. Leaving dreams unfulfilled is not the best way to leave this world!

28

The Corbett Foundation

The Corbett Foundation was established in 1994 by industrialist-turned-conservationist Dilip D. Khatau and his wife Mrs Rina Khatau. The idea of this new conservation organization came to them in the Corbett National Park, hence the somewhat 'restrictive' name, but now, 30 years after it was established, TCF functions in many parts of India, from Manipur to Assam through central India to Gujarat.

Coming back to Dilip Khatau, I first met him in 2001 at the Mumbai Golf Club, on the recommendation of Bittu Sahgal. That first meeting made us life-long friends, due to our mutual interest in conservation. Dilip facilitated my visit to the Naliya grassland in the Abdasa taluka of Kutch, and provided all help in his ancestral village, Tera. He was a fine host – friendly, compassionate and accommodative of different views. As soon as I retired in 2015, he asked me to become a scientific advisor in TCF, a position that I hold even now. He was a true leader – respected, admired and loved by all his staff – but at the same time he was strict when it came to results and performance. His death on 9 March 2023 left a void in the organization, but his planning ensured that TCF remained in good hands. Rina Khatau has taken over the mantle very well, so we all hope that Dilip's baby, TCF (now nearly 30 years old), will flourish for many more decades.

The Corbett Foundation has programmes in Kanha, Bandhavgarh, outside Kaziranga, Kutch and of course around Corbett. A small team of field biologists and social scientists are supervised by five senior scientists, each highly motivated and experienced. Kedar Gore, the director, is based out of TCF's headquarters in Mumbai, but he is mostly in the field. With more than 20 years of experience (with 10 years in WWF), Kedar is a highly respected conservationist. I first met him at a wildlife photo exhibition organized by WWF-India at the Max Mueller Bhawan, New Delhi, in 1998 during the Wildlife Week. The young Kedar was impressed with me, as I was meticulously reading and praising the captions of the photographs. Later he wrote to me, 'While most of the people just saw the pictures and moved on, you were reading and discussing the captions with great interest.' As Kedar lives in Mumbai, I met him many times after that; soft-spoken, slightly reserved and handsome, he leaves a good impression on anyone he meets.

A passionate birdwatcher since the age of 10, Kedar was drawn into the conservation world full-time in 1996. His overall area of work has been wildlife conservation, ecological restoration, human–wildlife conflict mitigation, biodiversity preservation, sustainable livelihoods, conservation education, advocacy and awareness. Kedar has extensively travelled across the length and breadth of India, covering more than 120 protected areas and wilderness landscapes. Since joining TCF, Kedar has been instrumental in launching and successfully implementing several landscape-level conservation programmes in Uttarakhand, Gujarat, Maharashtra, MP, Assam and Manipur states, which have benefitted various flagship species, such as the tiger, elephant, GIB, Lesser florican and Greater One-horned Rhinoceros, as well as local communities.

As scientific advisor to TCF, I work closely with Kedar and his senior staff. We have written two books together: *The Great Indian Bustard* (2016) and *Saving India's Wilderness: Challenges and Solutions* (2020). I regularly send good research papers and articles to Kedar and his senior team, while he too keeps me informed about all the major developments within TCF. We have been to the field many times together, in Kanha, Bandhavgarh, Corbett and Kutch; he is a good birdwatcher, better than me, and often points out minor differences between apparently similar-looking birds. I share his dream of making TCF the top conservation organization in India. Kedar is also on the governing board of BNHS. We both think that by leveraging each other's strengths, the two organizations, BNHS and TCF, can forge a powerful alliance – one where each organization's unique skills will complement the other perfectly.

Kedar has built a solid team; one of its core members is Devesh Gadhavi, who looks after the Gujarat programmes and the bustard/florican work. A connoisseur of Urdu literature, he always has an upperhand over me, as he remembers thousands of *shair*s and *ghazal*s, quoting them at the appropriate time during communication, while I hardly remember 10 *ghazal*s.

My only excuse (and not a very convincing one) is that I forget *shair*s and *ghazal*s, so every time I hear them, they seem new to me! Devesh works as the deputy director and senior programme officer (wildlife) at TCF. Our interest in floricans, other grassland species and bird photography binds our friendship. He has written 14 papers, many reports and popular articles in magazines. In 2021, he was appointed by the Hon'ble Supreme Court of India to be part of a three-member expert committee, which is looking into the prospect of putting powerlines underground in the GIB habitat.

Now, the biggest threat to the GIB is from the massive network of transmission lines that have come up in the Thar Desert. And not only in India – powerlines are one of the topmost threats to all bustard species across the world. As bustards have large beaks and their eyes are on the sides, they have poor frontal vision, due to which they cannot see the power or transmission lines and collide with them while flying. During the last nine years, 11–12 bustards have died after hitting the transmission lines. Therefore, the remaining small population of less than 100 birds have to be safeguarded from any additional mortality occurring due to the powerlines, which have contributed to more than 50 per cent of GIB deaths in the last 10 years. Not only GIB, but thousands of other birds, particularly eagles, storks and flamingoes, die after hitting the transmission wires or due to electrocution.

In June 2024, the expert committee constituted by the Supreme Court met the 'stakeholders', first in Jaisalmer and later at WII, Dehradun, where Devesh played a key role in supporting the recommendations of WII. I also attended the meeting as a 'stakeholder', and saw Devesh pleading with facts and figures for the protection of the GIB. The prime hope for the GIB in Rajasthan is a 13,000 sq. km expanse of landscape in Jaisalmer, comprising the 3,162 sq. km Desert National Park and the 2,807 sq. km Pokhran Field Firing Range (under the army) and the adjoining unprotected areas that hold more than 90 per cent of the bird's global population. The WII researchers, led by Dr Sutirtha Dutta, and the Rajasthan Forest Department have identified this as GIB Priority Area, where new transmission lines are to be avoided and the existing high-risk segments that are in the flight path of GIB – about 250 km in length, near the Desert National Park and Pokhran – are to be moved underground, whilst the rest of the transmission lines in the adjoining unprotected areas can be installed with good-quality Bird Flight Diverters. Research has shown that flight diverters (placed on the powerlines) can reduce bird mortality by 70–80 per cent. Unless the GoI takes strong steps and leaves the nearly 13,000 sq. km – what we call the GIB Arc – intact without new powerlines, the future of this grand bird is at stake. To mitigate climate change we need 'green energy' (solar panel farms, wind mills), but if this green energy results in the death of thousands of bird species and the extinction of the GIB, it cannot be called green energy.

Devesh works in the Kutch landscape, so he is TCF's bustard/florican and grassland expert. Much like the work done by Sujit Narwade of BNHS in the Thar Desert, Devesh takes up the issue of grassland species conservation of Kutch in national and international fora. He also represents TCF in various government committees. We discuss conservation issues regularly, and every six or eight months, I visit his field stations. Up to the early 2000s, there were 20–30 bustards in and around the Naliya grasslands – I have seen 9–11 displaying males – but thanks to the rapid development of wind mills and the network of high-tension wires in recent years, only two to three female bustards are left; there are no males, so technically the bustard is ecologically dead in Kutch. A few Lesser Floricans still visit the grasslands, but their number is also declining. We both are working to save the grasslands for other species, such as Chinkara, Indian Fox, White-browed Bushchat, Indian Courser, larks and warblers. Kutch is a dry region, with infrequent rainfall, and we are afraid that the uncertainties of climate change may adversely impact the remaining patches of the grasslands.

Another pillar of TCF is Dr Naveen Pande, a veterinary scientist who runs the Kaziranga programme. A good field naturalist, excellent communicator, superb administrator and a good writer, Naveen has an insatiable appetite to read – a habit I share and admire in him; his house is a veritable library. I first met Naveen at the BNHS field station at Bokakhat (Assam) in 2014, when I was supervising the Bengal Florican Project. In 2016, when I organized my first 'Scientific and Popular Writing Course' in Tera, Gujarat, for TCF scientists, Naveen was one of its most active participants. Whenever I meet him, he still talks about the course, which, he says 'left an indelible mark on me…changed my approach to […] work after I met you'.

Naveen is one of the strong pillars of TCF. We have done several field surveys in Assam and Manipur. Incidentally, our last field survey was in Manipur in March 2023, before the unfortunate violence broke out in May. Before that trip to Manipur, we spent a few wonderful days in Kaziranga, where his two young sons, Nalin and Shivam, also joined us. I wrote in my diary: 'The boys were desperate to see a wild elephant. To fulfil their wishes a majestic tusker appeared and provided excellent opportunities for photography, as it came close and crossed the road. It was the highlight of the trip.' Looking at the smiles on their young faces, I encouraged 14-year-old Nalin to write an article in *Saevus* magazine, which he did. When his article was published, Nalin wrote, 'I'm thrilled to find the digital edition of the *Saevus* magazine (June–August 2023). What a joy to see my article there! Your guidance and motivation made it possible. Many thanks, sir. Please suggest [another] topic and target site. I'll read more and then write.'

One of the most fulfilling aspects of the work that I do for TCF revolves around books. In September 2016, the IUCN held its World Conservation Congress in Hawaii, and in July of that year, Kedar Gore had asked me to write a book on the GIB, as TCF had included a discussion on the GIB as a side event of the Congress. As I knew the subject well, I wrote the first draft in two weeks, wrapping it up in mid-August. Amazing pictures were provided by Dhritiman and others, and within weeks, well ahead of Kedar's departure for the Congress, the book *Great Indian Bustard: A Pictorial Life History* was published. That is the fastest I have ever written a book!

In 2020, the second book was created to mark the 25 years of the founding of TCF. It was structured as 25 articles revolving around subjects or areas of interest to TCF. From early 2019 onwards, we started asking major conservationists to send in their articles. The result was *Saving India's Wilderness: Challenges and Solutions* (2020), funded by the Hem Chand Mahindra Foundation and TCF, and edited by me and Kedar Gore. Essays were contributed by experts such as Dr Johnsingh, Dr Renee Borges, Dr Ullas Karanth, Vivek Menon, Dr Aparajita Dutta, Ashok Mahindra, Dr Pankaj Sekhsaria, Dr Raghunandan Chundawat and others. This book, and the earlier one on the GIB, not only highlighted the extensive work done by TCF but also gave the organization significant visibility on a global scale.

One of the most successful programmes run by TCF is the Cattle Compensation Programme, funded by WWF-India. I consider it a remarkable on-the-ground conservation programme in India. A year after the establishment of this wonderful organization, Dilip Khatau realized the importance of cattle compensation and launched the Interim Relief Scheme in 1995, in and around the Corbett Tiger Reserve. The Interim Relief Scheme aims to provide immediate relief to owners whose cattle have been killed by large carnivores in and around the protected areas. World Wildlife Fund-India has been a partner in this scheme since 1997, and in 2016 the scheme was extended to Kanha Tiger Reserve in MP.

Most of our large protected areas are surrounded by extremely poor people. In an editorial titled 'Rich Biodiversity, Poor People' which appeared in the *Hornbill*,[55] I wrote: 'If we make a map of district/talukas with high poverty levels and superimpose it on the forest cover of India, the two will overlap, with few exceptions.' The tragedy is that poor villagers have to face the success of some tiger reserves in the form of increasing cattle kill by large carnivores, or crop damage by ungulates. Cattle compensation is given only if a tiger or leopard kills the animal outside the legal boundary of a park or tiger reserve. But even if cattle is killed outside, the legal process of getting compensation is so cumbersome (and corrupt) that some villagers resort to poisoning the carcass to get rid of the carnivore (most tiger/leopards return to eat the carcass, until the entire animal has been consumed). Here, the admirable role of the Interim Relief Scheme comes into play. As Dilip Khatau once told me: 'By providing immediate relief in less than 24 hours, the revenge killing of carnivores is avoided.' The Corbett Foundation staff also take up the case with the government for the final settlement of compensation. In the buffer zone of Kanha Tiger Reserve, the organization even places camera traps near livestock kills to monitor the big cats responsible for the kills. The data collected is shared with the Forest Department on a regular basis.

I have personally seen the workings of the Interim Relief Scheme in Corbett. It has transformed villagers' antagonism into whole-hearted support for long-term wildlife conservation goals. Under the Interim Relief Scheme, immediate ex gratia financial assistance is offered, supplementing the compensation that is later provided by the Forest Department for livestock losses. Over the years, around 20,000 livestock depredation cases from Corbett and Kanha have received interim ex gratia payments, without any recorded instances of retribution against big cats. The Interim Relief Scheme not only protects large carnivores but also fosters awareness about wildlife conservation through interactions with local communities during kill inspections. These interactions encourage active community participation in conservation efforts, and also provide crucial information for strategic decision-making, such as locations where livestock killing is taking place, the time when they typically happen as well the animal involved (tiger or leopard). Such basic information helps the forest department take precautionary steps to prevent further livestock killing. Because of the Interim Relief Scheme, many big cats have been saved from retaliatory action by aggrieved villagers, making it one of India's longest-running and most effective conservation programmes led by NGOs. This is a scheme worth replicating in all tiger reserves – we cannot have conservation at the cost of poor villagers.

Since TCF took up this programme in 1995, not even a single tiger or leopard has been poisoned. It is run by Harender Bargali, who looks after the Corbett programme of TCF. A seasoned conservationist (he was the co-chair of the Sloth Bear Expert Team in the IUCN/Bear Specialist Group for about 10 years), Harender also served as a member of the State Wildlife Advisory Board, Uttarakhand, during 2012–14. I see him as a key part of the younger crop of researchers and conservationists, who give me hope that the Indian conservation movement is in the right hands.

29

Looking Ahead

When I moved to Lucknow in 2018, it took me six to seven months to set up the house, but by the end of that year, I had in place a fully furnished house and a functional library.

When I had selected an apartment in the Eldeco Eden Park Apartments in 2008, the surrounding area had lots of trees and open spaces, but by the time I settled in Lucknow in 2018, the trees had been replaced by haphazardly constructed houses on three sides of the gated community. Fortunately, I chose an apartment at the rear, away from the main Kursi Road, so the noise of heavy traffic is not audible to me. But the houses visible from my seventh-floor back windows prove how not to develop a new colony. Up to the horizon, there are houses in every possible shape and size and colour, like a cheap jigsaw puzzle, with narrow streets, no place to plant trees, limited drainage facilities and no open space. Such haphazard housing colonies have been allowed to come up in a country that created Ajanta, Ellora, Meenakshi Temple, Taj Mahal, Shahjahanabad, Jaisalmer Fort, and landscaped cities like Mysore, Jaipur, Chandigarh and areas like Lutyens Delhi. I still cannot understand why (when we create new housing areas) we don't plan wide tree-lined roads, proper drainage, open parks, rows of houses and strategically located shopping centres. I can understand that nothing can be done in Old Lucknow, which was built for a population of 50,000 individuals. But when we build new housing colonies, particularly for middle-income people, what prevents us from planning properly?

Soon after I moved in, I found that the gated Eldeco Eden Park Apartments housed many working and retired scientists, and some of them became my friends, so I am never lonely. I have rediscovered the pleasure of reading books and writing articles. While I was doing my PhD or working in the field, I did not have much time to read books and articles. But after retirement, with no rush to go to work, I have plenty of time for my beloved books. It is like living the adolescent period of my life once more, when I had all the time in the world during long summer holidays to read books. But then, life is too short to read all the interesting biographies and books on history, geography, philosophy, literature, nature and wildlife.

My working life of nearly 50 years, when I read books connected with my research work, was sandwiched between two major reading periods: from mid-1960s till 1980, and then post retirement, from 2015 onwards. But I have always believed, and it's something I tell my students too, that one must read beyond one's subject of interest, especially books on arts, history, biographies, poetry in any language...anything; but one must read. Two years of Covid gave me sufficient time to re-read books from my own library. Some books, such as E.O. Wilson's *Sociobiology: The New Synthesis*, or Richard Dawkin's *Selfish Gene* can be read many times; so is the case with the wonderful books by Yuval Noah Harari. I am also a great fan of Arundhati Roy and have all her books and articles in my collection. Another favourite author is Ramachandra Guha. I feel proud that I know both of them personally and occasionally communicate with them.

I do not believe in buying books to decorate the drawing room, and can truthfully assert that I have read or used almost all the books that I have in my personal library. I do not believe in the beautiful Japanese word, *tsundoku*, which refers to the habit of acquiring books and letting them pile up at home without reading them.

I first read of a disease outbreak in China in November 2019. Soon, more news came in about the increasing number of deaths in a place called Wuhan. The name was unfamiliar, as I do not have much knowledge about the geography of China, except Tibet, the Yangtze River, the Taklamakan Desert, Outer Mongolia – all learnt from biogeography lessons. In early 2020, the world woke up to a looming Covid-19 pandemic, but it was in a distant land. In mid-February 2020, I attended the COP of Convention on Migratory Species at Gandhinagar, as a part of the UP Forest Department delegation. The following month, I went to Jaisalmer with Ashok Mahindra – former corporate honcho, wildlife photographer, writer, conservationist, philanthropist and founder of Hem Chand Mahindra Foundation – to see the situation of wildlife in the Desert National Park. There, on 2 March, we heard that an Italian tourist had tested positive for Covid, and so the hotel where he had stayed had been sealed. There was a pall of fear in the air, but we spent a wonderful week in the desert, thinking nothing would happen. On 20 March, I gave an interview to Lucknow Doordarshan on birds – things appeared normal but we were not expecting what followed next. On 22 March, a Sunday, a countrywide curfew was imposed as an experiment, followed by the 21-day lockdown on 24 March.

Soon, horror stories of the difficulties of poor people started to trickle in, keeping me sleepless. I was lucky to live in a gated colony on the seventh floor, and could afford food, water, electricity and appliances to keep myself in comfort, but what about the poor daily-wage earners? I decided that I had to do something and started giving money to the needy people, first in my own housing colony and then to the labourers who had worked in my house when I shifted to Lucknow. Fortunately, I had kept their mobile numbers.

It was a difficult period for every Indian, either financially or emotionally. Seeing the flight of poor people from big cities to their villages, walking hundreds of kilometres under the hot sun, was heartbreaking. Such scenes became 'normal' for the upper middle class, but not for me. Day by day, it became difficult to 'live life as usual' when my fellow citizens were on the streets, left hungry, thirsty and penniless by the callous government. Somehow, the first six months – the first wave of Covid – passed. When I heard that my friends, Ulhas Rane and Dr Parascharya, had died of Covid, the epidemic became personal. Fortunately, my BNHS family was safe, except for a few who got Covid but were recovering. Nonetheless, a bomb was ticking near me.

On 18 October, Neeraj Srivastava, whom I had not met for many months, came over for dinner. He had returned from an official tour of Bareily, and was unwell, not knowing that he had been infected with Covid. Within four days, I had high fever that refused to abate. First, I hid the fever from friends and relatives, and tried to treat myself at home, but when that didn't work I contacted the local Covid centre, which immediately sent doctors and medicines. They advised me to shift to the government hospital but I was reluctant, considering the condition of such hospitals. When my condition deteriorated further, there was no option, and on 30 October, I had to be rushed to a government-approved Covid hospital – Era's Lucknow Medical College and Hospital, started in 2001 by the Era Educational Trust. I stayed in the hermitically sealed Covid centre, with 50–60 other Covid patients, for seven days. At that time, there was no medicine for treating Covid, so the doctors and nurses, totally covered in personal protection equipment (PPE), fed us immunity-boosting medicines, food and juices.

The greatest help at this trying time was provided by my friend Sanjay Kumar, a high-ranking IAS officer. Before I was admitted to Era Hospital, Sanjay had called the CMO of Lucknow and the head of the hospital, which had ensured special treatment for me, something I was not aware of. As a result, I became suspicious when my photograph was taken by someone (in PPE apparel, so it was difficult to see the face) every day. When I finally asked a doctor, he said that they had instructions to report on my status to the hospital's CMO, who would then send the update to Sanjay!

Seeing two or three fellow patients die in the Covid room was truly depressing. Isolated as I was, sans books even, with only videos and podcasts on YouTube to keep me company, the highlight of my day was replying to the hundreds of WhatsApp messages that I would get from friends and acquaintances whom I had not contacted for many years.

It was only the love of my colleagues, relatives, students and friends that kept me alive when everything looked rather bleak.

I was discharged on 7 November, when tests showed that I was Covid-free. My nephew Obaid and assistant Sunny brought me home at around 9.00 p.m., but I could not recognize myself in the mirror – the shrivelled face, overgrown beard and dishevelled hair shocked me – but what was to follow was more dreadful.

Late that night, around 12.30 a.m., I started sinking. Somehow I opened the door, so that in case I died, people would not have to break the door! At 2 a.m., I reluctantly called my nephew and Sunny; they rushed me back to Era Hospital, 23 km away from my house. The hospital refused to admit me, as I was Covid-free. It took two hours for Era Hospital to tell Obaid to take me elsewhere, while my condition continued to deteriorate rapidly. Looking at my state, they told my nephew to take me to the nearest ICU. After driving for about a kilometre, my nephew found a small private hospital that was supposed to be a multi-speciality centre, but had limited facilities. At 4.30 a.m. there was no choice, so I was admitted to the small ICU centre that already had three patients. Though the doctors managed to stabilize me, the next eight hours were hell for me. So, after staying for one more night, I was shifted to Medox Hospital, about 2 km from my residence, where I stayed for another seven days. Dr Shailendra Singh, then head of the Turtle Survival Alliance, and his team were the ones who helped me find Medox Hospital, and provided a vehicle and other facilities.

As soon as Amjad, my assistant in Mumbai, heard about me, he rushed to Lucknow, and for the next three weeks he, along with Sunny, nursed me back to health. Shailendra and his team also ensured that I never felt alone.

The dedication of doctors and nurses, working tirelessly in their stifling plastic PPE gear with Covid-infected patients, would make any Indian proud. I salute them for saving hundreds of lives. As soon as the Covid vaccine became available, I got the injection at a primary health centre, which must have seen better days. One could make out that except for the Covid injection centre, everything else was in shambles, filthy or broken. Even in the poorest country of Africa, health centres would surely be better equipped? We have some of the finest hospitals in the world, with state-of-the-art medical facilities, but our primary health centres, where the majority of the population goes, lack basic facilities, medicines and equipment. Even clean toilets are not available. Some do not get 24-hour electricity, so they run on generators (that is, if the fuel is not usurped by the corrupt generator in-charge). I feel that India cannot develop unless we give emphasis to basic education and health. The fancy airports, jazzy malls, highways for rich people, fast trains for select few and publicity events cannot hide the underbelly of India. If the government has to give 5 kg rations every month to 80 crore Indians, there is something wrong in our development paradigm.

After retirement I executed three major projects that gave me the opportunity to continue my fieldwork, albeit to a lesser extent. The first was studying the movement pattern and dispersal of the Bengal Florican (*Houbaropsis bengalensis*). A satellite-telemetry-based pilot study, which started in 2013 and continued till 2017, the project entailed surveying the North Indian terai, including parts of Nepal and the Brahmaputra Valley. It was funded by the MoEF, and partly by the RSPB. We found that for almost six to seven months, the Bengal Florican moves out of the protected areas (e.g., Dudhwa, Pilibhit tiger reserves) and roams over grasslands and crop fields, where they are not safe from hunters and high-tension wires. One of our tagged birds died after hitting such a wire.

The second major project was on the threatened bird species of the Brahmaputra floodplains, funded by Swedish birder, conservationist and philanthropist, Per Undeland, who sponsored the India Bird Programme under BLI's Preventing Extinction Programme. It ran from 2015 to 2018, and helped highlight the plight of grassland-obligate species, such as Black-breasted Parrotbill (*Paradoxornis flavirostris*), Slender-billed Babbler (*Turdoides longirostris*), Jerdon's Babbler (*Chrysomma altirostre*), Manipur Bushquail (*Perdicula manipurensis*), Swamp Francolin (*Francolinus gularis*), White-throated Bushchat (*Saxicola insignis*), Jerdon's Bushchat (*Rhodophila jerdoni)*, Marsh Babbler (*Pellorneum palustre*), Swamp Prinia (*Prinia cinerascens*) and a few more. Ever since I saw those grasslands nearly 35 years ago, I have been wanting to unravel the mysteries of the birds found in those dynamic grasslands, which are dependent on heavy rainfall and annual flooding. So, this project gave me great satisfaction, for I was able to highlight the dire situation of the tall, wet grassland-dependent species, something no one had given attention to previously. Now, I am happy to know that a few students of Guwahati and Tezpur universities and scientists of NGO Aaranyak are currently working on these species.

My last project with BNHS was on the Sarus Crane (*Grus antingone*), spanning seven districts of UP. It started in 2016 and the final report was submitted in 2019. Funded by the UP Sarus Protection Society, it allowed me to extensively cover seven districts of UP, and study intensively the tallest flying bird in the world in its domain, the small wetlands of the Gangetic plains. I found that besides the conversion of small wetlands into crop fields or houses, there are two major threats: free-range stray dogs and network of high-tension wires. While the dogs eat the birds' eggs and chicks, the adults die after hitting the wires.

All three projects, a legacy of my long association with BNHS, resulted in final reports and papers, which we extensively distributed to the concerned authorities, so they could take conservation actions. As a scientific advisor to TCF, and informal supervisor to many young researchers, I still go to the field, particularly in Gujarat, Rajasthan, UP, Bihar, Assam and Jammu & Kashmir. I feel that reading and writing are fine, but the real action is in the grasslands of Kutch, the wetlands of UP, and the cool vales and mountains of Kashmir and Ladakh. The aerial display of the Lesser Florican in Sonkhaliya in Ajmer district, the unison call of a Sarus pair in Lakhimpur district, the fleeting sight of Markhor in the Limbar Wildlife Sanctuary in Kashmir or the unconcerned foraging of the Ibisbill in the Sindh River gives me energy and keeps me going.

Over the past 50 years of active wildlife research, I have witnessed profound changes in India's conservation efforts.

Although detailing these changes would require another book, I will summarize my experiences briefly. The pivotal moment was the enactment of the Indian Wildlife (Protection) Act in 1972, initiated by the then PM Indira Gandhi, who demonstrated unparalleled dedication to nature conservation. The only parallel I can think of is Sultan Qaboos of Oman, who was another notable leader in this regard.

Since India's independence, we have largely averted major species extinctions, although many, including 180 bird species, have seen population declines. Some species, such as the Mountain Quail, may have naturally dwindled, while others, such as the Pink-headed Duck, were hunted to extinction. The duck, once common in North India, is now reduced to about 80 museum specimens, as reminders of its former splendour in reed-covered wetlands.

There have been a few rediscoveries as well. The Jerdon's Courser was rediscovered by BNHS scientist Bharat Bhushan in 1986, leading to efforts to study and preserve its habitat, though the bird has gone missing again in the past decade. Similarly, the Forest Owlet, thought extinct for eight decades, was found in 1997. It is now known to exist in several protected areas across Maharashtra, Gujarat and MP.

The formation of the Indian Board of Wildlife in 1952 – aimed at developing national parks, sanctuaries and protecting wildlife – was a significant milestone. The board included prominent conservationists, such as Sálim Ali, E.P. Gee, M. Krishnan and R.S. Dharmakumarsinhji. Regular meetings were held under the leadership of the then PM Jawaharlal Nehru or senior ministers, highlighting the Board's importance, despite the nation's post-Partition challenges. Other government-sponsored programmes, such as the Project

Tiger, have been crucial in saving the majestic tiger from near extinction. I regard Project Tiger as one of the world's most successful conservation initiatives. Without it, the tiger's status in India would be dire.

Fifty years ago, BNHS was the major conservation organization, working all over India. Although there were other institutions, such as the Wildlife Preservation Society of India, established in 1958, and the much older Nilgiri Wildlife Association, established in 1877, these were more or less regional organizations. But now, we have more than a dozen major conservation organizations. Notable among them are WWF-India, WTI, Ashoka Trust for Research in Ecology and Environment, Aaranyak, Wildlife Protection Society of India, NCF, apart from numerous other local/regional organizations. Environment education, pioneered by the late Lavkumar Khachar in Gujarat and Dr Erach Bharucha in Maharashtra, has taken deep roots in college and university curriculums. The post-graduate course in wildlife, started by AVC College and AMU in the 1980s, is now taught in many universities.

What does the future of conservation in India look like? Recent successes in the breeding and reintroduction programmes for species such as the Pygmy Hog and vultures demonstrate the effectiveness of strategic conservation efforts. While efforts for the GIB and Lesser Florican are still in their infancy, their prospects for recovery are encouraging.

Species are resilient and can recover with proper administrative support, sufficient funding and long-term scientific conservation plans. Timely interventions have saved many species from extinction. For instance, by the late 1960s, only nine Mauritius Kestrels remained in Mauritius, with threats from habitat destruction, pesticides and invasive species. Once these threats were mitigated and breeding programmes initiated, their numbers increased to around 1,000 by 2019. Controlling threats such as pesticides, cats and monkeys was crucial. In India, however, the subject of control of stray dogs (particularly their elimination from protected areas), which threaten 80–90 endangered species, often sparks backlash from dog lovers.

A recent paper[56] in the journal *Science*, 'The Positive Impact of Conservation Action', proves that conservation actions are successful but need significant scaling-up to meet global targets. Successful conservation requires government support, long-term funding, science-based actions, public support and passionate commitment. The key pivot, however, is government backing, and all the successful examples I have discussed in this book had it – it is simply essential for conservation success.

As we confront modern challenges, we must ask: Do we have leaders today who embody such commitment to conservation? The plight of the GIB serves as a poignant reminder of the need for visionary leadership in safeguarding our natural heritage. It will be a shame for the country if we allow this grand bird to go extinct in the coming years. We know what to do to save it from extinction, but a firm commitment is required to protect its landscape habitat and to protect its surviving individuals from the network of high-tension wires, which now dominate its aerial habitat. The plight of the bustard epitomizes the threats, more or less similar, that many birds (such as cranes, storks, eagles and vultures) face daily. Signing biodiversity commitments in glittering international functions is fine, but the real work is on the ground, where some hard decisions have to be taken.

My great friend Aasheesh Pittie writes the following words under his email signature: *Quotlibros, quam breve tempus* (So many books, so little time). How very true! For lovers of books, Augustus' Latin quote is as valid now as it was 2,000 years ago. No one has time to read all the wonderful books that are published every year. Similarly, for the conservationist, one life is not enough to do all the conservation work that is required to keep this world habitable for all creatures, big and small. I have also fought the same fight, but whether I have succeeded or failed, only time will tell.

This is my life story, before the curtain falls.

Notes

Chapter 2

1. Rahmani, A.R. (1990), 'Precocious Display Behaviour in Birds', *Newsletter for Birdwatchers* 30 (9 & 10). p. 8.

Chapter 4

2. Rahmani, A.R. (1976), 'Should We Kill the Siberian Crane? Should We Destroy the Taj Mahal', *Imprint*, Bombay.
3. Rahmani, A.R. (12 February 1978), 'Threat to the Taj', *Patriot*.
4. Rahmani, A.R. (1 March 1979), 'Save Taj Mahal', *Concept*.
5. Islam, Z., Ugra, R. Rahmani, A.R., and Prakash, V. (1999), *Birds of Mathura Refinery*, Bombay Natural History Society and Indian Oil Corporation Ltd., Mumbai and New Delhi.
6. Rahmani, A.R. (1979), 'Narora – An Ideal Place for Bird Watchers', *Tourism and Wildlife*, Government of India, NOIDA.
7. Rahmani, A.R. (6 June 1977), 'Poachers of Narora', *National Herald*.
8. Rahmani, A.R. (1981), 'Narora Reservior, U.P.: A Potential Bird Sanctuary', *Journal of the Bombay Natural History Society* 78 (8892).
9. Rahmani, A.R. (1989), 'Narora P. Reservoir: An Excellent Habitat for Waterfowl', *Corsonat* (3 & 4). 79.

Chapter 6

10. Vardhan, H. and Goriup, P. (1980), 'Bustards in Decline', *Tourism and Wildlife Society of India*, Jaipur.
11. Rahmani, A.R. and Manakadan, R. (1985), 'Present Status of the Great Indian Bustard', *Bustard Studies* 3. pp. 123–131.
12. Rahmani, A.R. and Manakadan, R. (1986), 'Movement and Flock Composition of the Great Indian Bustard', *Journal of the Bombay Natural History Society* 83. pp. 17–31.
13. Rahmani, A.R. and Manakadan, R. (1987), 'Interspecific Behaviour of the Great Indian Bustard', *Journal of the Bombay Natural History Society* 84. pp. 317–331.
14. Rahmani, A.R. (1987), 'Protection to the Great Indian Bustard', *Oryx* 21 (3). pp. 174–179.
15. Rahmani, A.R. (1988), 'Conservation of the Great Indian Bustard in the Karera Bustard Sanctuary,' *Biological Conservation* 46. pp. 135–144.
16. Rahmani, A.R. and Manakadan, R. (1990), 'Past and Present Distribution of the Great Indian Bustard', *Journal of the Bombay Natural History Society* 87 (2). pp. 175-194.

Chapter 7

17. 'Looking Back at Stockholm 1972: What Indira Gandhi Said Half a Century Ago on Man and Environment.' *Down to Earth*, 29 May 2022, tinyurl.com/3yf7u63k. Accessed 26 September 2024.

Chapter 8

17. Sankaran, R., Rahmani, A.R. and Ganguli-Lachungpa, U. (1992), 'The distribution and Status of the Lesser Florican in the Indian Subcontinent', *Journal of the Bombay Natural History Society* 89. pp. 156–179.

18. Rahmani, A.R., Jha, R.R.S., Khongsai, N., Shinde, N., Talegaonkar, R. and Kalra, M. (2017), 'Studying Movement Pattern and Dispersal of the Bengal Florican (*Houbaropsis bengalensis*): A Satellite Telemetry Pilot Project. Final Report 2013–2016', Bombay Natural History Society. p. 157.
19. Rahmani, A.R., Ngulkholal Khongsai, N., Rahman, A., Imran, M., Taksh Sagwan, T. and Ojah, S. (2016), 'Conservation of Threatened Grassland Birds of the Brahmaputra Floodplains', Bombay Natural History Society. p. 66.

Chapter 9

20. Rahmani, A.R. (2019), 'The Trickster of the Desert: Greater Hoopoe Lark', RoundGlass Portal.
21. Rahmani, A.R., Shobrak, M. and Newton, S.S. (1994), 'Birds of the Tihama Coastal Plains of Saudi Arabia,' *OSME Bulletin* 32. p. 119.

Chapter 10

22. Rahmani, A.R. (1993), 'Little Known Bird: Whitebrowed Bushchat', *Oriental Bird Club Bulletin* 17. p. 830.
23. Rahmani, A.R. (1997), 'Status and Distribution of Whitebrowed Bushchat in India', *Forktail* 12 (1996). pp. 77–94.

Chapter 11

24. Daniel, J.C. (1970), 'The Tiger in India: An Enquiry – 1968–69', *Journal of the Bombay Natural History Society* 67(2).

Chapter 12

25. Islam, Z.A. and Rahmani, A.R. (2004), *Important Bird Areas in India: Priority Sites for Conservation*, IBCN, Bombay Natural History Society.

26. Islam, Z.A. and Rahmani, A.R. (2008), *Potential and Existing Ramsar Sites in India*, Oxford University Press India.
27. Rahmani, A.R. and Islam, Z. (2008), *Ducks, Geese, and Swans of India: Their Status and Distribution*, IBCN.
28. Rahmani, A.R. (2011), *Threatened Birds of India Their Conservation Requirements*, Oxford University Press India.

Chapter 13

29. Salim Ali (1981) President's Letter: To change or not to change? This is the question. *Hornbill*, 1981 (4): 5-6.
30. News (1992) Rare Book Exhibition, *Hornbill* vol. 1 (January-March), page 21.

Chapter 17

31. News (1992) Rare Book Exhibition, *Hornbill* vol. 1 (January-March), page 21.

Chapter 18

32. Rahmani, A.R. (1996), 'Status of Vultures in the Thar Desert of India', *Vulture News*. 35. p. 23.
33. Rahmani, A.R., (1998), 'Decline of Vultures in India', *Newsletter for Birdwatchers* 38 (5). pp. 80–81.
34. Prakash, V. (1999), 'Status of Vultures in Keoladeo National Park, Bharatpur, Rajasthan, with Special Reference to Population Crash in *Gyps* Species', *Journal of the Bombay Natural History Society* 96. pp. 365–378.
35. Green, R.E., Newton, I., Shultz, S., Cunningham, A.A., Gilbet, M., Pain, D.J. and Prakash, V. (2004), 'Diclofenac Poisoning as a Cause of Vulture Population Declines Across the Indian Subcontinent', *Journal of Applied Ecology* 41. pp. 793–800.

Chapter 20

36. Rahmani, A.R. (1990), 'Distribution, Density, Group Size and Conservation of the Indian Gazelle or Chinkara (*Gazella bennetti*) in Rajasthan', *Biological Conservation* 51. pp. 177–189.
37. Rahmani, A.R. (1990), 'Distribution of the Indian Gazelle or Chinkara in India,' *Mammalia* 54(4). pp. 605–619.
38. Rahmani, A.R. and Sankaran, R. (1991), 'Blackbuck and Chinkara in Rajasthan: A Changing Scenario', *Journal of Arid Environments* 20. pp. 379–391.
39. Rahmani, A.R. (1989), 'Status of the Blacknecked Stork (*Ephippiorhynchus asiaticus*) in the Indian Subcontinent', *Forktail* 5. pp. 99–110.
40. Rahmani, A.R., Narayan, G. and Rosalind, L. (1990), 'Status survey of the Greater Adjutant *Leptoptilos dubius* in the Indian Subcontinent', *Colonial Waterbirds* 13(2). pp. 139–142.
41. Javed, S., Qureshi, Q. and Rahmani, A.R. (1999), 'Conservation Status and Distribution of Swamp Francolin in India', *Journal of the Bombay Natural History Society* 96. pp. 16–23.
42. Iqubal, P., McGowan, P.J.K., Carroll, J.P. and Rahmani, A.R. (2003), 'Home Range Size, Habitat Use and Nesting Success of the Swamp Francolin (*Francolinus gularis*) on Agricultural Land in Northern India', *Bird Conservation International* 13. pp. 127–138.
43. Haq, I.U., Rahmani, A.R., Bhat, B.A., Ahmad, K. and Rehman, S. (2022), 'Breeding Biology of Ibisbill (*Ibidorhyncha struthersii*) in the Kashmir Himalayan Region of India,' *Waterbirds* 44(3). pp. 356–362.
44. Haq, I.U., Rahmani, A.R., Bhat, A.B., Rehman, S., Ahmad, K. and Ahmad, R. (2022), 'Disturbances to Ibisbill (*ibidorhyncha struthersii*; Vigors, 1832) in Sindh Valley, Kashmir Himalaya,

India', *Journal of the Bombay Natural History Society* 119.
45. Tiwari, J.K. and Rahmani, A.R. (1997), 'The Current Status and Biology of the White-naped Tit (*Parus nuchalis*) in Kutch, Gujarat, India', *Forktail* 12(1996). pp. 79–102.
46. Ali, S., (1955). The birds of Gujarat. Part II. *JBNHS* 52 (4): 735–802. (see page 785).
47. Zarri, A.A. and Rahmani, A.R. (2004), 'Wintering Records, Ecology and Behaviour of Kashmir Flycatcher (*Ficedula subrubra*)', *Journal of the Bombay Natural History Society* 101(2). pp. 262–268.

Chapter 21

48. Jegannathan, P., Green, R.E., Bowden, C.G.R, Norris, K., Pain, D. and Rahmani, A.R. (2002), 'Use of Tracking Strips and Automatic Cameras for Detecting Critically Endangered Jerdon's Coursers (*Rhinoptilus bitorquatus*) in Scrub Jungle of Andhra Pradesh, India', *Oryx*, 36 (2). pp. 182–188.
49. Andhra Pradesh Forest Department (2010), 'A Species Recovery Plan for Jerdon's Courser (*Rhinoptilus bitorquatus*)', Andhra Pradesh Forest Department, Government of Andhra Pradesh, Hyderabad. pp. 1–30.
50. Rasmussen, P.C. and Collar, N.J. (1999), 'On the Hybrid Status of Rothschild's Parakeet (*Psittacula intermedia*; Aves, Psittacidae)', *Bulletin of the Natural History Museum London* (Zoology) 65. pp. 31–50.
51. Ali, S., and Crook, J.H. (1959), 'Observations on Finn's Baya (*Ploceus megarhynchus* Hume) Rediscovered in the Kumaon Terai', *Journal of the Bombay Natural History Society* 56 (3). pp. 457–483.

Chapter 23

52. Rahmani, A.R., Kumar, B., Rahman, F., Sohail, M., and Mehta P. (2022), 'Sarus Crane: Life in Human-Dominated Landscape', *International Journal of Ecology and Environmental Sciences* 48. pp. 673–686.

Chapter 24

53. Rahmani, A.R. (1976), 'Birdwatching in Kukrail', *The Pioneer*, Lucknow.

Chapter 28

54. Rahmani, A.R. (2013), 'Rich Biodiversity: Poor People', *Hornbill* April–June. pp. 1–2.

Chapter 29

55. Langhammer, Penny F. *et al.* (2024), 'The Positive Impact of Conservation Action', *Science* 384 (6694). pp. 453–458.

Acknowledgements

Firstly, I want to acknowledge the support of my deceased parents, my sisters and brothers who allowed me to lead my life the way I wanted. Secondly, I thank the BNHS family for making me what I am today. There are too many names to acknowledge but I will not do justice if I do not mention Mr J.C. Daniel, my initial boss and then my colleague, for mentoring and guiding me throughout my career. Dr Salim Ali, Prof. A.H. Musavi, Mr B.G. Deshmukh, Ms Dilnavaz Variava, Dr Pheroza Godrej, Dr Ashok Kothari, Mr Bittu Sahgal, Mr Debi Goenka are some people who need special mention.

Among BNHS staff, I am particularly grateful to Sachin Kulkarni, Gopi Naidu, Dr Raju Kasambe, Dr Zafarul Islam, Noor Khan, Dr Gayatri Ugra, and Isaac Kehimkar. Lack of space does not allow me to mention all the wonderful people with whom I have worked.

I am grateful to all my students for tolerating my exactitudes, and often idiosyncratic behaviour.

I am thankful to the previous and present directors of BNHS, Dr Bivash Pande and Mr Kishor Rithe respectively, for allowing me to consult old files while writing my life story.

I shall be failing in my duty if I do not mention a few close friends: B.C. Choudhury, Nalini Choudhury, Goutam Narayan, Dhritiman Mukherjee, Raghunandan Chundawat, Joanna van Gruisen, Anwaruddin Choudhury, Neeraj Srivastava, Ashish Chandola, Sanjay Kumar, Qamar Qureshi and Parikshit Goutam.

Lastly, I want to thank Anita Mani for her wonderful editing and asking incisive questions that vastly improved the text. She is a great editor. My final thanks to the team at Juggernaut for putting the book together.

A Note on the Author

Dr Asad R. Rahmani joined the Bombay Natural History Society (BNHS) in 1980, where he worked as a senior scientist for 12 years before joining the Department of Wildlife Sciences, Aligarh Muslim University, in 1991. He rejoined BNHS as its director in mid-1997 and led the organization until his retirement in 2015. During his tenure at BNHS he was part of several committees of the Ministry of Environment and Forests, Government of India. Dr Rahmani is a Scientific Consultant to The Corbett Foundation and Hem Chandra Mahindra Trust. He was on the Board of Wetlands International South Asia (WISA) till 2022 and is currently on the Governing Council of BNHS.

Dr Rahmani's main interest is in grassland and wetland birds, with a focus on highlighting the plight of lesser-known species and habitats. He has published 26 books, 150 peer-reviewed scientific papers, 90 book reviews, nearly 75 editorials and nearly 400 popular articles on nature conservation. He was a Global Council member of BirdLife International, UK (2006–2013), and chairman, BirdLife Asia Council (2006–2013). Dr Rahmani was the executive editor of the *Journal of the Bombay Natural History Society*

(JBNHS) from 2005 to 2015. He was accorded lifetime achievement awards by Vasundhara Abhiyan, Pune, and Nature Mates, Kolkata, and named Member of Honour by BirdLife International, UK.

About Indian Pitta

Indian Pitta is India's first dedicated book imprint for bird and nature lovers, conservationists and policymakers. Our books about birds and natural history go beyond field/identification guides, to explore the bigger mosaic of habitats, ecosystems and human interactions that touch the lives of birds and animals. Successful conservation programmes, troubling environmental challenges, personal exploration of a landscape, deep dives into the ecology of a species, the quest for a rare species and the sheer joy of birding – these are some of the ideas that you can expect to explore within the pages of our books.

Also Available

ISBN: 978-81-959969-0-2
Price: 599/-

ISBN: 978-93-5345-181-3
Price: 499/-

ISBN: 978-93-5345-665-8
Price: 599/-